Acknowledgments

Improved Recovery was researched and written by Nicholas J. Constant, a contract writer for Petroleum Extension Service.

The American Petroleum Institute Advisory Committee for the School of Production Technology of Petroleum Extension Service was active in determining content coverage and in reviewing the manuscript. The following members were particularly helpful:

Marvin C. Boyd, Sun Exploration and Production Company

R. G. Carson, Exxon Company, U.S.A.

Danny Echols, ARCO Oil and Gas Company

J. W. Hargis, Mobil Producing Texas and New Mexico, Inc.

S. H. Stocker, Amoco Production Company

Special thanks for content review are due to Randy Schlotz, Texas Pacific Oil Company; E. F. Herbeck, Atlantic Richfield Company; J. Del Broomell, ARCO Oil and Gas Company; Rodney R. Maxson, Sun Exploration and Production Company; and Ralph Long, Getty Oil Company.

Content consultation was also provided by John D. Morelli, Department of Zoology, and by Royal E. Collins and Robert S. Schechter, Department of Petroleum Engineering, The University of Texas at Austin.

In-house review was provided by consultants Bruce Whalen, Ron Baker, and Dick Donnelly.

How to Use This Manual

The format of this manual includes a set of specific objectives for each section; at the end of the section is a competency self-test. To get maximum benefit from the manual, read the specific objectives carefully before studying the material in each section. The self-test is based on the objectives. As you study the material in the section, take notes, using the objectives as a guide to the most important parts.

When you feel that you have mastered the objectives, begin the self-test. Since it is a self-test, *you* must decide whether you should refer back to the material to answer the questions by determining how important that section is to your work. If you feel that you need to be very competent in an area, do not refer back until you have finished the test. This way, using the scoring points given at the beginning of the test, you can determine your percentage of competency. Score the test by using the corresponding key provided at the end of the manual.

Introduction

Immense quantities of known oil remain untapped in the United States. Of the 460 billion barrels of oil discovered before the early 1980s, government and industry studies estimate that just over one-fourth has been produced (fig. 1). Of the remaining 339 billion barrels, only 27 billion can be recovered by conventional primary and secondary recovery methods. Recovery efficiencies in U.S. oil fields average about 19 percent for primary recovery and about 14 percent for secondary recovery.

Various methods have been developed to recover part of the residual oil. They are often referred to as *enhanced oil recovery* (EOR). However, the difference of opinion about the precise meaning of this term can lead to confusion when it is used. Therefore, this manual will refer to methods of recovering residual oil as *improved recovery*.

Waterflooding has been the main type of improved recovery for several decades. Fields that respond well to waterflooding may yield much greater percentages than the 14 percent average for secondary recovery. However, even reservoirs that perform best under waterflooding leave behind a sizable portion of oil. Moreover, nearly all the candidates have already been identified and are being waterflooded. One study predicts that conventional secondary production in known fields will decline from 50 percent of U.S. production in 1980 to below 10 percent by 1995. The same study expects that the use of advanced recovery methods such as chemical flooding, miscible gas injection, and steam drive will then account for over 75 percent of U.S. production, excluding Alaska.

The government estimates that advanced methods, in their current state of development, can recover up to another 50 billion barrels. The remaining 250 billion barrels are a target for recovery by future technologies. A comparison may help convey the magnitude of 50 billion barrels of oil; this quantity is roughly equal to five new Alaskan oil fields the size of Prudhoe Bay or one-third of Saudi Arabia's oil reserves.

In the 1980s, the oil industry is undergoing a transition from primary reliance on waterflooding to the increasing use of advanced recovery methods. These methods are more expensive than waterflooding. They use injection fluids and heat that are more costly than water. They often require specialized production equipment. And they incur high capital costs due to long delays between the initial investment and the sale of the produced oil. The cost effectiveness of using these methods depends on a complex interplay of economics, government policy, and technological advancement. In mid-1980, the cost of recovering 1 barrel of oil by these methods ranged from—

$22 to $46 for chemical flooding;

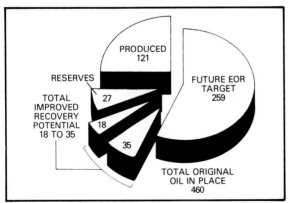

Figure 1. Domestic oil resources available for improved recovery, in billions of barrels. (Reprinted with the permission of *Well Servicing* and the Association of Oilwell Servicing Contractors)

1

$26 to $39 for miscible gas injection (carbon dioxide); and

$21 to $36 for thermal recovery (steam and in situ combustion).

These figures include a 15 percent return on investment. When the price of oil was around $30 per barrel, advanced methods could be applied economically to the more promising oil fields (fig. 2). Before the price increases of the 1970s, however, much of the remaining known U.S. oil was not economically recoverable.

A continued increase in the real price of oil would naturally encourage wider application of advanced recovery methods, but the real price cannot increase indefinitely. One reason is that price increases encourage conservation. By the early 1980s, lower demand, due to conservation coupled with worldwide economic slowdowns, had produced a slight oversupply. Few analysts believed the oversupply was permanent, but some thought that the price of oil would

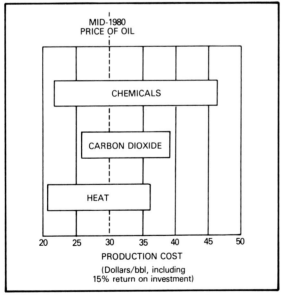

Figure 2. Range of production costs for improved recovery methods in mid-1980. Source: Lewin and Associates, Inc., *Economics of Enhanced Oil Recovery, Final Report.* DOE/ET/12072-2. Washington, D.C.: U.S. Department of Energy, 1981.

not increase any faster than the rate of inflation, at least until the mid-1980s. Furthermore, even as oil prices rise, the prices of oil-derived chemicals and fuel oil used in improved recovery operations go up in parallel, and the economic benefits of improved recovery methods may be lost.

Government policy also has mixed effects on the use of improved recovery. On the one hand, some price controls were lifted in the late 1970s and early 1980s. On the other hand, government financial aid in the form of cost sharing and favorable tax treatment for research and pilot projects may decrease during the 1980s. The Energy Research Advisory Board, a federal policymaking board, recommended in 1982 that the government decrease its sponsorship of energy research and development. The board did, however, give higher priority to improved recovery than to other sources of energy such as coal liquefaction.

The most certain boost to improved recovery may come from technological development. New technologies can lower production costs, increase efficiency, make production easier, and increase supplies of needed materials. For example, more economical ways to produce expensive injection chemicals will allow them to be used more widely and intensively. Genetic engineering may be used to develop bacteria that will produce oil-recovery chemicals when they come in contact with subsurface oil. Another possibility is the development of better ways to control the movement of injected substances so that they contact more of the remaining oil.

Regardless of these uncertainties, it is likely that the remaining oil will be needed at some future point, and the imperative of this need will bring about favorable circumstances for the use of advanced improved recovery methods.

1

Petroleum Reservoirs

OBJECTIVES

Upon completion of this section, the student will be able to:

1. Identify and define three characteristics of reservoirs.

2. Identify two main types of rock found in reservoirs, and describe factors that influence their permeability and porosity.

3. Identify and define types of traps.

4. Describe the contents of reservoirs and the natural pressures operating in them.

5. Identify methods of evaluating reservoirs for improved recovery projects and the data they provide.

6. Describe ways data are organized to gain a complete description of reservoir structures and fluids.

INTRODUCTION

Improved recovery, the production of the hard-to-get portion of reservoir oil, is a challenge requiring detailed knowledge of underground conditions. For this reason, petroleum industry personnel who want to understand improved recovery methods must have an introductory knowledge of reservoir characteristics, the behavior of reservoir fluids, and the methods used to gather and analyze information about individual reservoirs.

Important reservoir characteristics to understand are porosity and permeability, types of reservoir rocks and their traits, types of reservoir structures, types and distribution of reservoir fluids, and natural sources of pressure in reservoirs. Additional information on fluid behavior in reservoirs is provided in section 3, "Waterflooding." It is also helpful to be familiar with the methods of gathering and organizing information and to appreciate the importance of using a conceptual model of a reservoir as a basis for making operational decisions during an improved recovery project.

RESERVOIR CHARACTERISTICS

Oil is believed to be the product of organic material from plant and animal life, which was trapped or buried in silt. As more and more layers of silt were deposited over millions of years of time, this organic material was slowly transformed into oil by bacteria, heat, pressure, and chemical action. The beds of silt in which oil had formed became compacted under the pressure of overlying layers, and oil and gas squeezed out into adjacent rocks that were less susceptible to compaction. Oil continued to migrate through rock formations until it reached the surface of the earth or was trapped in rocks by some type of barrier.

Rock formations in which oil accumulates are potential reservoirs from which oil can be produced. In order for a formation to qualify as an economically profitable reservoir, it must meet three conditions:

1. The reservoir rock must have enough porosity to contain profitable amounts of oil.
2. The reservoir rock must have enough permeability to allow the oil to flow through it to wells.
3. The reservoir must have a barrier that prevents the oil from escaping and allows it to accumulate.

A petroleum reservoir, then, is a rock capable of collecting, containing, and giving up oil and gas. To be commercially productive, the reservoir must be a rock mass with enough pore space to contain a worthwhile amount of hydrocarbons. The rock must also give up the contained fluids at a satisfactory rate when penetrated by a well.

Porosity and Permeability

Close examination of a rock with a powerful magnifying glass reveals openings or pores. A rock with pores is said to be porous or to have porosity. The *porosity* of a formation is its capacity to contain reservoir fluids, and is expressed as a percentage of total rock volume. The greater a rock's porosity, the more fluids it can hold. Porosity may vary from less than 5 percent in a tightly cemented sandstone or carbonate to more than 30 percent for unconsolidated sandstones. Accurate determination of the porosity of a rock formation is difficult.

Besides porosity, a reservoir rock must also have *permeability;* that is, the pores of the rock must be connected. These passages between pores allow petroleum to move from one pore to another. The rock's permeability determines how easy or difficult it is for the oil to move or flow within the rock. When a well is drilled into a permeable formation, the petroleum has a way to flow out of the pores and into the well.

Permeability is measured in darcys. A porous medium has a permeability of 1 darcy when a pressure drop of 1 atmosphere across a sample 1 cm long and 1 cm² in cross section will force a liquid of 1-cp viscosity through the sample at a rate of 1 cm³ per second. Most petroleum reservoirs have permeabilities so small that they are measured in thousandths of a darcy, or millidarcys (md). The permeability of consolidated sandstones ranges from a few to several hundred millidarcys. Unconsolidated sandstones and carbonates with vugs or fractures often have permeabilities in the thousands of millidarcys. The permeability of carbonates with small-diameter channels, however, can be less than one millidarcy. The relationship between the porosity and the permeability of a formation is not necessarily direct. A rock may be highly porous but have low permeability. However, high porosity often coincides with high permeability.

Table 1 lists some combinations of porosity and permeability in actual oil field sands. The values shown for the East Texas field show that these characteristics vary even

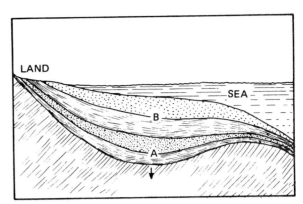

Figure 3. The weight of overlying unconsolidated sediments (B) compacting a sediment layer (A) into sedimentary rock

within different portions of the same kind of rock.

Reservoir Rocks

Most oil-bearing formations are made of *sedimentary rock,* which was formed from decayed organic matter or eroded rock particles. These materials were carried to their present location by wind or water and deposited in layers, or strata. Most sedimentary rocks have been formed in the ocean. The earth is continually being eroded by water and wind, and the land is being worn down and washed out to sea (fig. 3).

Two types of sedimentary rocks, sandstones and carbonates, are the most common reservoir rocks. *Sandstone* is rock formed from compacted sand and coarse silt. *Carbonate* rock includes limestone and

TABLE 1
Porosity and Permeability of Selected Oil Field Formations

Horizon	State	Field	Porosity (Percentage)	Permeability (Millidarcies)
Woodbine	Texas	East Texas	22.1	3,390.00
Woodbine	Texas	East Texas	19.7	192.00
Wilcox	Oklahoma	Oklahoma City	16.9	677.00
Gloyd Lime	Louisiana	Rodessa	20.0	130.00
San Andres Lime	Texas	Goldsmith	12.0	50.00
San Andres Dolomite	Texas	Levelland	10.1	5–10

5

its relative, dolomite. *Limestone* is formed in the ocean by coral reefs or beds of shells. *Dolomite* is formed when part of the calcium in limestone is replaced by magnesium, a process known as *dolomitization*.

The porosity of sandstones and carbonates may be determined by the shape and closeness of the rock grains or by chemical or physical changes such as dolomitization, the formation of solution channels, or fracturing. Porosity may also be reduced by *cementation*—the solidification of minerals in rock pore spaces; or by *compaction*—the shrinkage of rock from the weight of overlying rock layers, evaporation, or other causes.

Sandstones. In sandstones, porosity is determined by the sizes and shapes of grains, which affect the way the grains are packed together. Porosity is greatest when grains are spherical and all one size, but it decreases as grains become more angular, because such grains pack together more closely. Spherical grains pack together in two ways (fig. 4). Rock with open or cubic packing (*A*) has a porosity of about 48 percent. Rock with rhombohedral packing (*B*) has a porosity of only about 26 percent because the grains are packed into a smaller

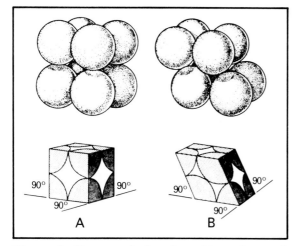

Figure 4. Open and closed packing of round grains of rock. *A*, cubic; *B*, rhombohedral.

space. Artificially mixed clean sand has a porosity of about 43 percent for extremely well sorted sand, grain size having little effect. Porosity decreases to about 25 percent for poorly sorted medium- to coarse-grained sand. Very fine-grained sand still has over 30 percent porosity.

Compaction by weight of the overburden, or overlying material, squeezes sand grains closer together; at greater depths this force may crush and fracture the grains. The result is smaller pores and lower porosity and, more importantly, a drastic decrease in permeability. A sandstone reservoir that can produce oil at a depth of 10,000 feet may become much too impermeable to be of any economic value at 20,000 feet. Cementation also tends to increase with depth.

Carbonates. The porosity and permeability of carbonates are determined by the way the sediment was deposited and by the changes that took place after the rock was formed. The original sediments formed lime crust, mud, or sand. These sediments were altered by cementation, compaction, leaching by water that moved through the rock, and dolomitization. When lime is converted into dolomite, particles may dissolve and disperse, leaving large holes called *vugs*. The presence of vugs increases porosity, but there may be little or no permeability if the passages between the vugs are small or absent. Another type of dolomite is formed of small crystals; the grains are fine and a large network of passages between the pore spaces creates high permeability. Such rocks often yield profitable volumes of oil.

Traps

Reservoirs are rock formations that trap and accumulate oil migrating from the source beds. These *traps* all have the same elements; a mass of porous permeable rock and an impermeable barrier to stop further oil movement. Although the elements are the same, the shapes and structures of traps vary depending on the way in which they were formed by the accumulation of sedimentary rock layers, earth movements, and erosion. Common types of traps are structural folds, faults, plugs, unconformities, and permeability changes. Combinations of these types may occur in reservoirs, an example being a faulted anticlinal trap.

Structural folds include *anticlinal* and *dome traps;* in these reservoirs, oil accumulates in the upfolded arches of rock layers. The oil moves in from underlying rock and migrates upward until it is stopped by a cap rock, a surface layer of impermeable rock (fig. 5). The structures are called anticlinal traps if the upfold is long and narrow and dome traps if the upfold is circular.

In *fault traps,* rock layers are broken by the pressure of an earth movement, and one

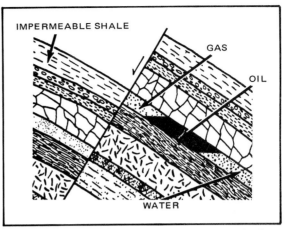

Figure 6. Fault trap. The block on the right has moved up from the block on the left, moving impermeable shale opposite the oil-bearing formation. (Courtesy of the American Petroleum Institute)

side of the break moves up, down, or sideways from the other side of the break, forming a *fault.* As a result, the broken end of the oil-bearing rock layer comes into contact with an impermeable rock layer such as shale that acts as a seal (fig. 6).

In *plug traps,* great masses of fluid rock push up into other rock formations. The *salt-dome trap* is formed when a salt plug pushes upward and breaks into overlying rock, forming a seal to the surrounding layers of oil-bearing rock (fig. 7). Oil may

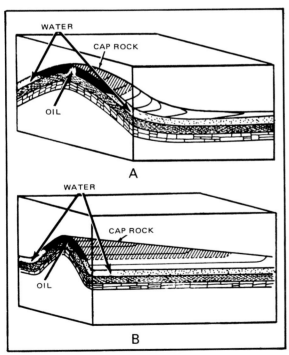

Figure 5. Structural traps with upfolded rock strata. *A*, dome; *B*, anticline. (Courtesy of the American Petroleum Institute)

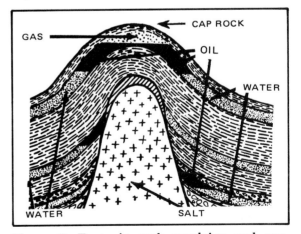

Figure 7. Traps in rock overlying and surrounding an impermeable salt dome. (Courtesy of the American Petroleum Institute)

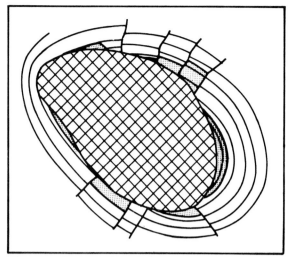

Figure 8. Scattered peripheral traps surrounding a piercement salt dome.

accumulate in many separate traps around a salt dome, closed off against the dome or against faults (fig. 8). The scattered arrangement of oil accumulations makes salt-dome traps difficult to drill economically. The geologist knows that the traps are there, but he cannot accurately predict their precise locations. Another type of plug trap, the "serpentine" plug, serves as a reservoir rather than a barrier (fig. 9). Serpentine

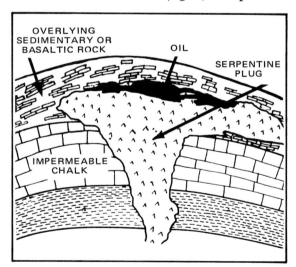

Figure 9. Reservoir in a porous serpentine plug that has intruded into surrounding impermeable formations. (Courtesy of the American Petroleum Institute)

plugs, named for the dark green rock sometimes found in them, are actually cones of volcanic debris filling the craters of old submarine volcanoes. The debris was formed into rock by chemical reaction with water and air and by compaction. The plugs are surrounded by impermeable chalk and may be covered with newer layers of sedimentary rock, basaltic rock from later volcanic eruptions, or limestone reefs. Oil may be found within the plugs themselves or within the newer layers of rock covering them.

In *unconformities,* land surfaces are eroded, exposing part of the underlying rock layers. Later, these exposed layers are covered by a new layer of impermeable rock that forms a seal (fig. 10).

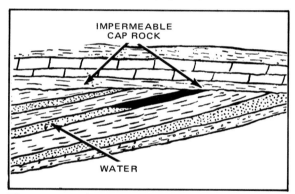

Figure 10. Oil trapped under an unconformity. (Courtesy of the American Petroleum Institute)

Lenticular or *lens* traps (so named because their shape resembles a double-convex lens) pinch out from the center to a thin edge all around. They are sealed at the top and sides by a sudden decrease in the permeability of the rock (fig. 11). Oil is held in the porous parts of the rock by the surrounding non-porous parts. Lenticular traps are formed by the irregular depositing of sand and shale. They also occur in the upper part of limestone reefs.

8

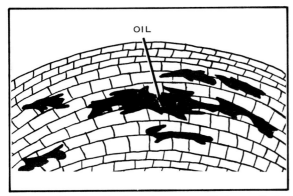

OIL

Figure 11. Lenticular traps confining oil in permeable parts of the rock. (Courtesy of the American Petroleum Institute)

Fluid Content and Distribution

Petroleum reservoirs usually contain varying proportions of three fluids—water, oil, and gas. These reservoir fluids are confined under pressure. When a well penetrates the reservoir, the difference in pressure between the reservoir and the bottom of the well causes the fluids to flow toward the well. Gravity may also aid the flow.

Water. Reservoirs are composed of sediments that were deposited on the seafloor. Originally, these sedimentary beds were completely saturated with salt water. Later, petroleum displaced part of this water. Salt water remaining in the formation is called connate or interstitial water. *Connate,* from the Latin meaning *born with,* refers to the fact that the water was in the formation when development of the reservoir began. *Interstitial* refers to the fact that the water is found in the interstices or pore connections of the formation.

Invariably, some connate water is distributed throughout the reservoir. In addition, nearly all petroleum reservoirs contain *free water* that supplies the water drive in certain reservoirs. *Bottom water* is free water located beneath the oil accumula-

tion; *edge water* is free water occurring at the edge of the oil-water contact zone.

Oil. Oil is lighter than water and does not mix with it. Oil makes room for itself in the pores of the reservoir rock by pushing water downward. However, not all of the water is displaced. A film of connate water clings to the rock surrounding the pore spaces.

Gas. Natural gas is often associated with reservoir oil and water in two forms, as solution gas and as free gas in gas caps. Virtually all oil contains some dissolved gas, or *solution gas,* when discovered. The amount of solution gas depends on the temperature, the pressure, the oil composition, and the amount of available gas. When the oil is brought to the surface and the pressure is relieved, much of this gas comes out of solution and the oil volume shrinks. Thus, reservoir engineers must take account of the fact that a barrel of oil in the reservoir will produce less than a barrel of oil in the stock tank.

Free gas—gas that is undissolved in oil—tends to accumulate in the highest portion of a reservoir and to form a gas cap. As long as the gas cap remains, the oil in the reservoir will be saturated with solution gas. Gas in solution lowers the oil's *viscosity,* or resistance to flow, making the oil easier to move to the wellbore.

Contact zones. In reservoirs containing oil and free water, contact between the two fluids occurs in a zone of part water and part oil that may vary from near zero to several feet thick. In gas-cap reservoirs, a similar contact zone exists between the oil and the gas cap. The lower the permeability of the oil-bearing formation and the greater the difference in the densities of the oil and the water, the thicker is the oil-water contact zone. The gas-oil contact zone tends to be narrower than the oil-water contact zone.

9

Reservoir Pressure

The pressure in most reservoirs is caused solely by the fluids they contain. At one time, the aquifers (water-bearing rocks) of these formations had fluid contact with the surface. Therefore, their fluid pressure is about equal to the pressure of a column of salt water at the same depth—a little less than ½ pound per square inch (psi) for every foot of depth. In some reservoirs, however, formation pressure is abnormally high. In these formations, overlying rocks were forced to move downward by earth stresses that occurred after the connection of the aquifers to the surface was severed. In these reservoirs, pressure is caused not only by reservoir fluids but also by the weight of the overlying rock, which can press additional water from adjacent formations into oil-bearing sandstone.

Reservoir pressure, also called *formation pressure*, causes the flow of fluids toward a production well drilled into the reservoir. If a pressure difference is created between the reservoir and the wellbore, fluids move in the direction of the well. If the formation pressure is high enough, it not only moves fluid to the wellbore but also overcomes the weight of the column of fluids, raising them to the surface.

DESCRIBING RESERVOIRS

A thorough geologic description of a reservoir's structure and content is essential in planning and conducting improved recovery operations. The target is oil that nature does not give up easily, and the strategy is to introduce heat or fluids that will mobilize the remaining oil. To decide whether a project is worth undertaking, engineers must first find out how much oil remains. Then if the prospect seems encouraging, they have

to make decisions about which improved recovery method will work best, where to locate injection wells, how much heat or fluid to use, and for how long.

The data needed to answer such questions are obtained primarily from well logs, core samples, and formation tests. Additional information may be obtained from seismic tests and strat tests. These data are organized into contour maps, isopach maps, and cross sections that describe the reservoir and its fluid contents in detail.

Evaluation Methods

Well logs. Well log data are used to determine formation characteristics such as rock types, porosity, permeability, the location of potential pay zones, and fluid saturations. These characteristics are influential in deciding the feasibility of a project, the choice of an improved recovery method, and the selection of well completion and production equipment. In evaluating depleted reservoirs for potential improved recovery projects, it is essential to determine the residual oil saturation with a margin of error of no more than 2 to 3 percent, as saturation is a key factor in deciding whether or not a project will be economically feasible.

The following types of logs may be used in evaluating a reservoir:
1. drilling-time logs;
2. mud logs;
4. electric well logs;
5. radioactivity well logs; and
6. acoustic or sonic well logs.

Certain logs, such as mud logs, provide direct measurements of properties of interest. Most well logs, however, measure another physical property that is then used to determine the desired property by applying known relationships between the two properties under certain reservoir conditions. Some logs provide quantitative measures; others indicate only qualities of the rock formation.

A combination of logs is selected based on the conditions of a particular reservoir. Data from the various logs are compared with each other and with data from nearby wells to arrive at the most comprehensive and accurate results possible. Comparing data enables analysts to overcome the limits of each measure and to reduce the margin of error.

Drilling-time logs provide an initial record of the depth, thickness, and relative hardness of formation beds; they give the first general picture of a well and provide a basis for later decisions.

Mud logs are used to detect the presence of oil and gas in the drilling fluid. Bit cuttings collected from drilling fluid help to identify rock types, and can give indications of permeability and porosity.

Electric well logs measure spontaneous and man-made electrical currents in rock formations. They are used in improved recovery projects mainly to calculate residual oil saturation, but they can also be used to determine other formation characteristics. The spontaneous potential log measures natural electrochemical activity between drilling and formation fluids of differing salinity, and is used to detect permeable beds and locate boundaries between beds. Resistivity logs measure the resistance of a formation to the flow of electricity induced by a magnetic field or sent down the wellbore from the surface. Of these, lateral focus logs and induction logs are used to calculate fluid saturations; normal logs are used to identify the boundaries of thin rock beds; and microspaced resistivity logs (a combination of normal and lateral focus

logs) are used to identify permeable intervals. In improved recovery analyses, the log-inject-log procedure is often used with resistivity logs to increase the accuracy of oil saturation measurements. A resistivity log is run on the formation in its original state. Next, hydrocarbons are displaced from the formation by injecting formation water into the pores. Finally, another resistivity log is run on the water-filled formation. Comparison of the two sets of data allows a determination of hydrocarbon content. In addition to resistivity logs, *dielectric constant logs* are used to detect oil, water, and oil-water mixtures. These logs measure the differing ability of water and oil to store electrical potential under the influence of an electric field.

Radioactivity well logs measure patterns of natural and induced radioactivity in subsurface formations. These patterns are used to determine porosity, rock types, and the presence of particular fluids and chemical elements in a formation. Types of radioactivity logs include gamma ray logs, neutron logs, and density logs. *Gamma ray logs* are used to distinguish between impermeable and permeable rocks, to detect natural fracture systems, and to distinguish between swelling and nonswelling clays. *Neutron logs* are used to determine porosity and to detect the presence of chemical elements such as hydrogen, carbon, oxygen, chlorine, silicon, and calcium. From these data, analysts can determine water and hydrocarbon saturations, salt content, and rock types. *Density logs* measure the bulk density of the rock matrix, fluids, and pore space. They are used to calculate porosity and to get an indication of hydrocarbon density.

Acoustic or sonic well logs measure the differences in travel times of sound pulses sent through formation beds. These data are used to determine rock types and porosity.

Core samples. *Core samples* are cylindrical samples of rock specially cut from reservoir zones of interest by a core bit and recovered in a cylindrical core barrel. They are used to determine rock types, porosity, permeability, and fluid saturations. Cores can be tested in the laboratory under conditions that simulate reservoir pressures and temperatures to determine the rate at which fluids will flow through formation rock to the wellbore and to evaluate the effect of a specific improved recovery method such as chemical flooding on fluid movement. Additional tests may be used to determine oil gravity, the salinity of formation water, and the mineral composition and microscopic structure of the rock. Although expert cutting, transporting, and testing of cores are costly and time-consuming chores, core analysis can reveal a great amount about reservoir rock characteristics.

Formation tests. Formation tests measure patterns of pressure in a formation and provide data useful in deciding whether and how to use a well. Two testing methods of interest in planning improved recovery projects are the drill stem test and the pressure test.

The drill stem test (DST) is a temporary well completion used to sample formation fluids and to determine whether the well will fulfill its intended purpose. The DST can

1. identify the fluid content of oil-bearing strata;
2. measure the average effective permeability of the surrounding formation to the fluid being produced;
3. detect the presence of wellbore damage and measure the resulting loss of permeability;
4. indicate the presence of impermeable structures and their approximate distance from the wellbore; and

5. determine the rate of flow into the wellbore under known conditions.

When a DST is run, a set of packers and valves is lowered into the wellbore on drill pipe. The packers seal off the test area from the mud-filled annulus. Specially perforated drill pipe allows reservoir fluids to flow into and up the drill stem. After a period of flow, the valves are closed to shut in the formation so that pressure will build up to formation-pressure level in the wellbore. A test usually includes two periods of flow and shut-in. The first round adjusts wellbore conditions so that better data can be obtained on the second round. A record of pressures is obtained during the test sequence.

Several improved recovery methods involve pumping fluid down injection wells; as the fluid spreads through the formation, it drives oil and gas toward adjacent production wells. *Pressure tests* can be run on completed wells to measure pressure changes resulting from changes in injection or production rates. Several types of tests are used, depending on the information desired. The production or injection rate of a well can be changed, and the resulting pressure changes in one or more adjacent wells can be measured. Pressure buildup can be measured in shut-in production wells or injection wells being started up. Pressure decline can be measured when a shut-in well begins to produce or pumping is stopped at an injection well.

Pressure tests provide the same kinds of information as drill stem tests: fluid content, rate of flow, permeability of the surrounding formation, the presence of impermeable structures, and the presence of wellbore damage. Test results may indicate that different completion equipment is needed to maximize production or that well stimulation is needed to overcome loss of permeability due to formation damage. Pressure tests also show how fluids travel through the reservoir; for example, they can indicate whether an injection fluid is following a desired pattern of flow. By showing the degree of fluid communication between injection and production wells, pressure tests allow an estimate of the future relationship between injection pressure and rate of flow. They provide information useful in planning improved recovery projects. They cannot be used with thermal recovery methods, however, because the calculations used to interpret the data assume a constant temperature throughout the reservoir.

Seismic data. *Geophysical exploration* is the measurement of the physical properties of the earth in order to locate commercial accumulations of oil, gas, or other minerals; obtain information for the design of surface structures; or make other practical applications. One of the most important properties studied in the oil industry is seismic vibration. Sensitive instruments are used to measure subsurface sound patterns that may yield information about possible oil-bearing structures. Data about these structures may be useful not only in locating reservoirs but also in planning improved recovery projects later on.

The seismic method employs the *seismograph* to measure the transit time of sound waves through the earth. This time depends on the nature of the rock being penetrated. Time can be measured for a wave reflected directly from the *shot point* or point of origin to the detector, or for a wave refracted indirectly through a subsurface rock layer (fig. 12).

When a seismograph party looks for subsurface structures, sound waves are sent down through the earth and return from the

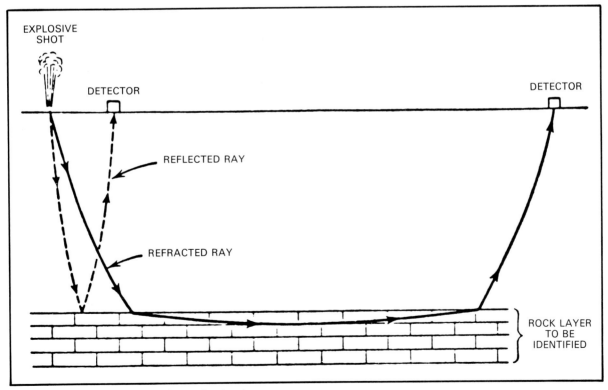

Figure 12. **Reflected and refracted seismic waves**

13

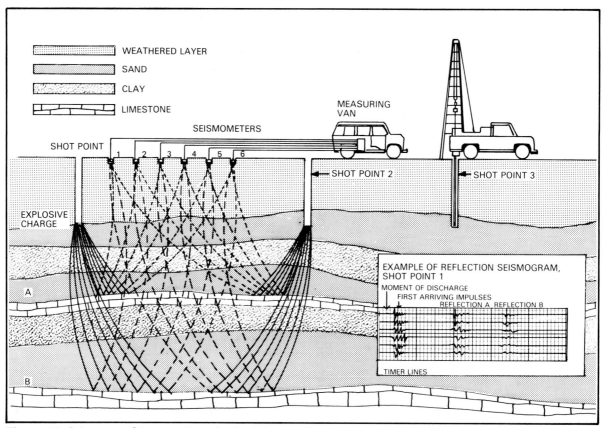

Figure 13. Seismographic mapping of subsurface geology

layers of rock (fig. 13). Geophones at the surface pick up the sound echoes and convert them to electrical impulses, which are recorded on a seismogram. Under favorable conditions, a geologic bed can be mapped quite accurately.

Strat tests. *Stratigraphy* is the study of the origin, composition, distribution, and succession of rock strata, which are distinct, generally parallel beds of rock. A stratigraphic test, commonly called a *strat test,* involves drilling a well primarily to obtain geological information. Data from several wells can be correlated to form a picture of subsurface structures of a formation (fig. 14). Strat tests are generally made during the exploration stage, but the information they provide can be useful later on for planning improved recovery projects.

Organizing the Data

Maps. Data obtained from the examination and correlation of logs can be used to prepare structural contour maps, isopach maps, and cross sections. *Contour* maps consist of a number of contours or lines that connect points of equal value, such as elevation above or below sea level. Geologists use structural contour maps to show subsurface features (fig. 15). Contour lines indicating elevation are drawn at regular depth intervals to enable geologists to depict three-dimensional shapes clearly. Contour maps can also be made to show the direction of faults and their intersections with beds and other faults; porosity; permeability; and structural arrangements such as old shorelines, pinch-outs, and truncated beds.

14

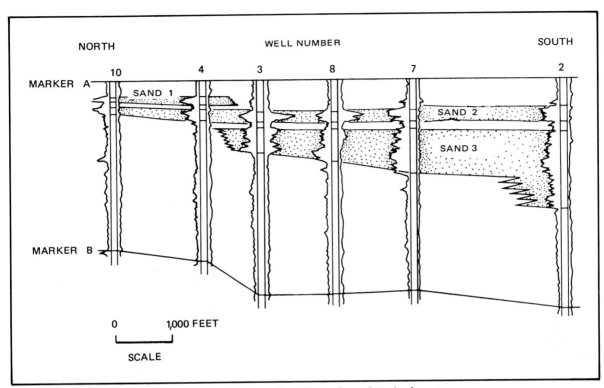

Figure 14. Stratigraphic cross section showing the effect of *sand 3* pinch-out

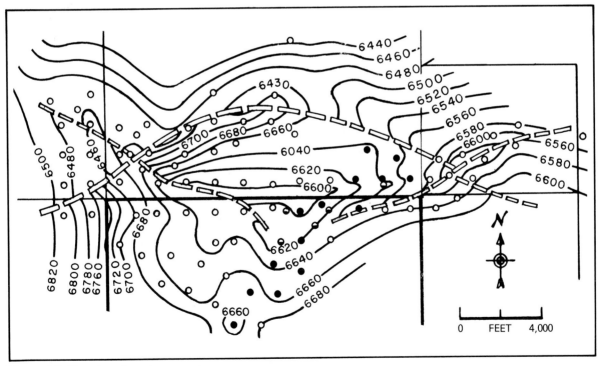

Figure 15. Structural contour map

15

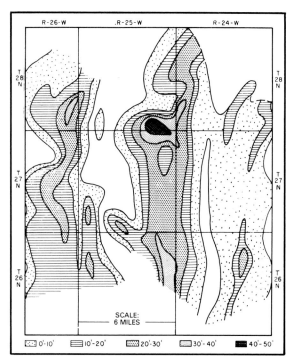

Figure 16. Isopach map of a formation with a length of 15 to 20 miles and a thickness of up to 50 feet

Isopach maps show variations in the thickness of formations (fig. 16). They are widely used in planning improved recovery projects.

Maps, like the floor plan of a house, give a view of subsurface formations from directly overhead. To supplement the information provided by maps, engineers and geologists use *cross sections,* which depict features of vertical sections of the earth. Figure 14 is an example of a cross section; it shows how *sand 3* extends from *well 2* through *wells 7* and *8* to *well 3* but is pinched out (thinned out) before reaching *well 4.*

Conceptual models. From the data gathered and correlated by various methods, geologists develop a *conceptual model* of a geologic area — an idea of the way the area is structured and the most likely locations of petroleum accumulations. A model's prediction of an oil-bearing formation's areal

extent; trend; and variations of porosity, permeability, and thickness are useful in making operational decisions during both primary and improved recovery. Without an accurate model during primary development, a well drilled a few hundred feet from a good producer may fail to find the target sand. "Missing" sand is likely to be even more disconcerting when improved recovery is attempted.

To appreciate the value of a conceptual model, consider the example of an Illinois oil field where secondary recovery by waterflooding was started before an accurate model had been developed. The results were puzzling: waterflooding worked in

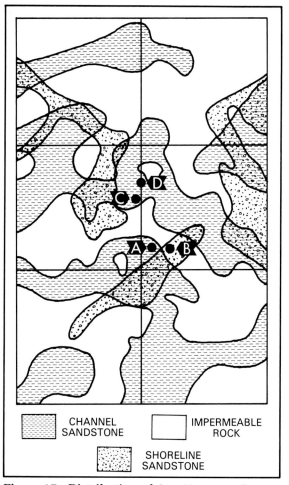

Figure 17. Distribution of Aux Vases sandstone reservoirs

16

some instances but not in others. These differences were explained by the model worked out by engineers. It showed that the field was composed of two different types of sandstone, shoreline stone and channel deposits (fig. 17). Lack of fluid communication between the two types of stone caused many of the problems. For example, production *well A* was located in channel sandstone. Injection *well B* was located nearby, but it failed to increase the production of *well A* because it was drilled into shoreline sandstone. By contrast, the production of *well D* was increased by waterflooding from nearby injection *well C.* Both of these wells were located in channel sandstone. The model aided engineers in selecting locations for additional wells by showing the thickness as well as the location of the two sandstones (not shown in fig. 17).

Self-Test

1: Petroleum Reservoirs

(Multiply each correct answer by four to arrive at your percentage of competency.)

Fill in the Blank and
Multiple Choice with One or More Answers

1. Define the following terms:

    (1) porosity _____

    _____

    (2) fault _____

    _____

    (3) connate water _____

    _____

    (4) permeability _____

    _____

2. In order to be an effective reservoir, a rock formation must meet the following three conditions:

    (1) _____

    _____

    (2) _____

    _____

    (3) _____

    _____

3. The two most common types of sedimentary reservoir rocks are (1) _____ and (2) _____.

4. _____ The porosity of a formation may be reduced by –
    A. dolomitization.
    B. compaction.
    C. cementation.
    D. open-packed spherical grains.
    E. all of the above.

5. Salt-dome traps are often difficult to drill because _____

_____

_____

6. In lenticular traps, oil is prevented from escaping by_____

_____

_____.

7. Gas dissolved in reservoir oil is called _____.

8. When a reservoir contains zones of oil, free gas, and free water, in what order are they located, from top to bottom?

    (1) top: _____

    (2) middle: _____

    (3) bottom: _____

9. In most reservoirs, the formation pressure that forces oil toward the wellbore is exerted mainly by _____.

10. In evaluating improved recovery projects, it is essential to determine the residual oil saturation with a small margin of error, because _____

_____

_____.

11. _____ The rate at which fluids will flow through formation rock to the wellbore can be determined by –

    A. laboratory testing of cores.
    B. acoustic well logs.
    C. drilling-time logs.
    D. the log-inject-log procedure.
    E. all of the above.

12. Well log data are used to determine formation characteristics such as

    (1) _____ and

    (2) _____.

13. Formation tests can detect changes in _____ in a formation.

14. _____ The physical features of a vertical slice of the earth can be represented on –
   A. an isopach map.
   B. a topographical map.
   C. a contour map.
   D. a cross section.
   E. all of the above.

15. What is a conceptual model of a reservoir, and how does it help engineers to plan an improved recovery operation? _____

_____

_____

_____

_____

2

Production Energies in Oil Recovery

21

INTRODUCTION

The first period in the producing life of a reservoir is called *primary production* or *primary recovery* (table 2). During this stage, natural reservoir energies displace oil from the pores of a formation and drive the oil toward production wells and up to the surface. These energies are provided by the expansion of liquid or gas contained in the reservoir; the encroachment of water into oil-bearing rock from adjacent aquifers (water-bearing rock); and the action of gravity.

Before or after the depletion of natural drives, engineers and owners may consider the technical and economic feasibility of *improved recovery*, the introduction of artificial drive mechanisms into reservoirs in order to recover a portion of the remaining oil. The purpose of improved recovery is the maintenance or restoration of formation pressure and fluid flow in a substantial portion of a reservoir. The energy is usually introduced through injection wells located in rock that has fluid communication with the production wells.

During primary or improved recovery operations, production may be increased by artificial lift and well stimulation methods. Artificial lift methods raise oil to the surface by pumping or gas lift. Well stimulation methods increase the permeability of the rock surrounding a production well by acidizing or fracturing the rock.

PRIMARY RECOVERY

Primary recovery is the initial stage of oil recovery in which natural reservoir energies are used to produce oil. These energies are known as *reservoir drive mechanisms* and include gas or depletion drive, water drive, and gravity drainage. Often, oil is produced by a combination of these drives.

Gas Drive

Gas drive, also called depletion drive, operates in closed reservoirs in which water contact with reservoir hydrocarbons is minimal. The expansion of compressed or dissolved gas moves the oil to the wellbore and toward the surface. Two kinds of gas drive are solution-gas drive and gas-cap drive.

Solution gas has provided the most common natural source of energy for oil production. In a reservoir with *solution-gas drive* (also known as *dissolved-gas drive*), the lighter hydrocarbon components exist as liquids in the reservoir at formation

TABLE 2
Types of Oil Recovery

Stages of Recovery	Methods of Recovery
PRIMARY RECOVERY	1. **Natural reservoir drives** Water drive Gas (depletion) drive Gravity drainage Combination drive
IMPROVED RECOVERY	2. **Waterflooding** 3. **Immiscible gas injection** Natural gas Flue gas Nitrogen 4. **Miscible gas injection** Carbon dioxide Hydrocarbons (propane, methane, enriched methane) Nitrogen 5. **Chemical flooding** Polymer flooding Micellar-polymer flooding Alkaline (caustic) flooding 6. **Thermal recovery** Steam drive Steam soak In situ combustion

pressures above the *bubble point*, the pressure at which gas begins to come out of solution. The oil is compacted under pressure. As production begins, pressure at the bottom of the well drops below the original formation pressure, allowing the compacted oil to expand and overflow into the wellbore. As the pressure goes below the bubble point, dissolved gas begins to expand and come out of solution, displacing more oil and forcing it into the wellbore (fig. 18). The pressure drop spreads further away from the wellbore, and gas continues to come out of solution and to displace oil. In such reservoirs, pressure declines rapidly and continuously, and wells generally require pumping or some form of pressure maintenance at an early stage. The *gas-oil*

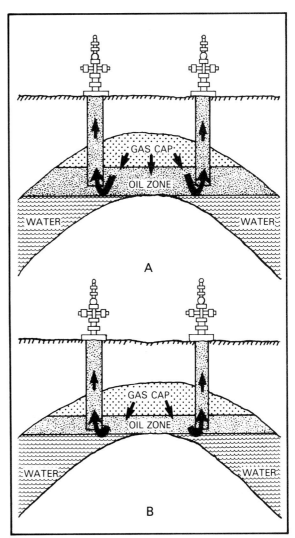

Figure 19. Gas-cap drive reservoir. *A*, original conditions; *B*, partially depleted.

ratio (the amount of gas compared to the amount of oil being produced) is initially low, then rises to a maximum and drops. With the exception of edge wells that have penetrated the oil-water contact, little or no water is produced. *Recovery efficiency* (the percentage of oil in place that is produced) varies from 5 to 30 percent.

In a reservoir with a *gas cap*, the free gas provides energy. As oil is withdrawn and pressure in the oil zone declines, the gas cap expands and pushes oil out of the pores (fig. 19). The pressure of the gas cap on the

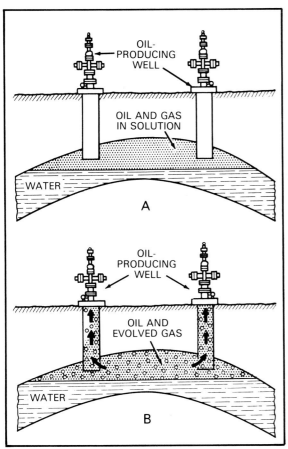

Figure 18. Solution-gas drive reservoir. *A*, original conditions; *B*, partially depleted.

23

oil zone helps to slow the expansion of solution gas and thus delays the loss of pressure resulting from its depletion. Gas escaping upward from the oil zone joins the gas cap and helps to maintain its pressure. Pressure decline may be slower than that of solution-gas drive when the production rate is properly controlled and the gas cap is maintained as a well-defined zone. Gas-oil ratios rise continuously in upstructure wells with this drive, and little or no water is produced in wells that penetrate the reservoir near its edge. Flow life depends on the size of the gas cap, and oil recovery may be from 20 to 40 percent of the original oil in place.

Water Drive

Water drive occurs when water in the reservoir provides enough energy to move oil out of the reservoir, into the wellbore, and part or all of the way to the surface (fig. 20). The water in most water-bearing formations has a fluid pressure proportional to the depth beneath the surface; in other words, the deeper the water, the higher the pressure.

As the oil is driven from the reservoir, the water moves in to replace it. This process is similar to emptying a tank of oil by injecting water into the bottom of the tank. The pressure remains high as long as the volume of withdrawn oil is replaced by an approximately equivalent volume of water. If the reservoir pressure remains high, the surface gas-oil ratio remains low because little or no free gas is evolved in the reservoir. Because a high reservoir pressure is maintained, wells usually flow on their own until water production becomes excessive and kills the well. Water production may begin early and increase to an appreciable amount as water encroaches into the oil and into the production wells. The expected oil recovery with water drive is generally higher than gas

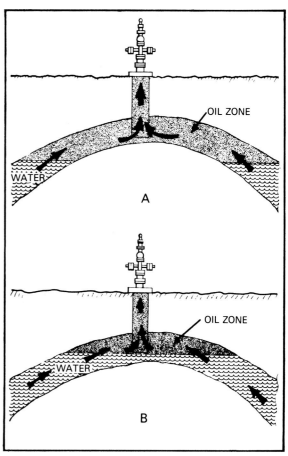

Figure 20. Water-drive reservoir. *A*, original conditions; *B*, partially depleted.

drive—up to 50 percent or more of the oil originally in place—because of the greater displacement efficiency of water.

Water-drive reservoirs can have bottom-water or edgewater drive. In a reservoir with *bottom-water drive*, the oil zone is underlain by a water zone. A well drilled anywhere through such a reservoir penetrates oil first and then water. In a reservoir with *edgewater drive*, the oil accumulation almost completely fills the reservoir. Water occurs only on the edges of the reservoir. Wells drilled along the edges penetrate water. Wells drilled near the top of the structure penetrate only oil. Bottom-water drive is usually less efficient than edgewater drive. More water tends to enter wells in bottom-drive reservoirs; as a result, less oil is usually produced.

Gravity Drainage

The force of gravity is always at work in a reservoir. Although the pressures exerted by rising water or expanding gas are usually the dominant natural drives in producing oil, gravity may also aid in production. As the original gas or water drives in deeper fields decline, gravity becomes a more important force. In certain shallow, very permeable reservoirs with steep underground dips, gravity drainage may be the dominant means of producing fluids.

Combination Drive

Gas or water drive can operate as the only drive mechanism in a reservoir. However, many reservoirs have a combination of drives. For example, gas-cap drive may be combined with water drive when a reservoir contains a water zone; both types of energy drive oil into the well. Another combination is gas-cap drive and gravity drainage. Free gas exerts pressure downward and gravity exerts an additional downward force. Solution gas may also work together with bottom water.

IMPROVED RECOVERY

For many reasons, a reservoir may produce only a small fraction of the oil in place during primary recovery. Production of the reservoir may have been started before good development and production practices were known. Unforeseen problems such as a casing leak or a blowout may have wasted reservoir energy, or the owner may not have been willing to invest more money for pressure maintenance while wells were producing profitably. However, even with the best primary production methods, a great deal of oil is left behind.

The purpose of improved recovery is restoration of formation pressure and fluid flow to a substantial portion of a reservoir in order to recover as much of the remaining oil as possible. There are four major methods of improved recovery: waterflooding, gas injection, chemical flooding, and thermal recovery.

Waterflooding, the most commonly used method, is often begun during primary recovery to prevent the decline of formation pressure as natural reservoir drives are depleted. It may be used to boost natural water drive or to replace natural gas drive. Gas is also used to maintain or replace pressure in reservoirs with natural gas drive. For this purpose, it is generally injected under relatively low pressures and is immiscible (does not mix) with the oil. Thus, this type of improved recovery is called immiscible gas injection. The gas used is generally natural gas, but other gases such as nitrogen or flue gas may be used.

The introduction to this manual refers to miscible gas injection, chemical flooding, and thermal recovery as advanced recovery methods because they not only restore formation pressure, but they also improve oil displacement by overcoming reservoir forces that would otherwise keep oil trapped in rock pores, or they improve patterns of fluid movement through the reservoir. In miscible gas injection, injected gas mixes with reservoir oil and releases it from the rock pores. The gas may be naturally miscible with oil or may become miscible under high pressure and other favorable reservoir conditions. Gases used for miscible injection include methane, methane enriched with other light hydrocarbons, propane, carbon dioxide, and nitrogen. In chemical flooding, detergentlike chemicals

TABLE 3
Improved Recovery Methods

Method of Recovery		Process	Use
WATERFLOODING	Water	Water is pumped into the reservoir through injection wells to force oil toward production wells.	Method most widely used in secondary recovery.
IMMISCIBLE GAS INJECTION	Natural gas, flue gas, nitrogen	Gas is injected to maintain formation pressure, to slow the rate of decline of natural reservoir drives, and sometimes to enhance gravity drainage.	Secondary recovery.
MISCIBLE GAS INJECTION	Carbon dioxide	Under pressure, carbon dioxide becomes miscible with oil, vaporizes hydrocarbons, and enables oil to flow more freely. Often followed by injection of water.	Secondary recovery or tertiary recovery following waterflooding. Considered especially applicable to West Texas reserves because of carbon dioxide supplies located within a feasible distance.
	Hydrocarbons (propane, high-pressure methane, enriched methane)	Either naturally or under pressure, hydrocarbons are miscible with oil. May be followed by injection of gas or water.	Secondary or tertiary recovery. Supply is limited and price is high because of market demand.
	Nitrogen	Under high pressure, nitrogen can be used to displace oil miscibily.	Secondary or tertiary recovery.
CHEMICAL FLOODING	Polymer	Water thickened with polymers is used to aid waterflooding by improving fluid-flow patterns.	Used during secondary recovery to aid other processes during tertiary recovery.
	Micellar-polymer (surfactant-polymer)	A solution of detergentlike chemicals miscible with oil is injected into the reservoir. Water thickened with polymers may be used to move the solution through the reservoir.	Almost always used during tertiary recovery after secondary recovery by waterflooding.
	Alkaline (caustic)	Less expensive alkaline chemicals are injected and react with certain types of crude oil to form a chemical miscible with oil.	May be used with polymer. Has been used for tertiary recovery after secondary recovery by waterflooding or polymer flooding.
THERMAL RECOVERY	Steam drive	Steam is injected continuously into heavy-oil reservoirs to drive the oil toward production wells.	Primary recovery. Secondary recovery when oil is too viscous for waterflooding. Tertiary recovery after secondary recovery by waterflooding or steam soak.
	Steam soak	Steam is injected into the production well and allowed to spread during a shut-in soak period. The steam heats heavy oil in the surrounding formation and allows it to flow into the well.	Used during primary or secondary production.
	In situ combustion	Part of the oil in the reservoir is set on fire, and compressed air is injected to keep it burning. Gases and heat advance through the formation, moving the oil toward the production wells.	Used with heavy-oil reservoirs during primary recovery when oil is too viscous to flow under normal reservoir conditions. Used with thinner oils during tertiary recovery.

and water are injected into the reservoir to decrease surface tension, break up the oil into small droplets, and carry it out of rock pores. Polymer chemicals are used to thicken injected water and thus to increase the sweep or coverage of the flood through the reservoir. Thermal methods are used to recover thick, heavy oils by heating them and causing them to flow more freely.

Three terms often heard in discussions about improved recovery are *secondary recovery, tertiary recovery,* and *enhanced oil recovery.* Because of the dynamic growth of the field of improved recovery, these terms have been used with more than one meaning. In reading literature or talking with others about improved recovery, it is helpful to be alert to the way these terms are being defined. The term *secondary recovery* has been nearly synonymous with water-flooding and immiscible gas injection because of the widespread use of these methods to maintain reservoir pressure. The term *tertiary recovery* has referred to methods that improve oil displacement and fluid movement in the reservoir in addition to restoring reservoir pressure. However, as the field develops, this distinction between two groups of sequentially used recovery methods is not holding true. For example, in fields with heavy thick crude oils, thermal methods such as steam drive may be used as the first or the second method of recovering oil because the methods ordinarily used don't work. Most improved recovery methods, in fact, can be used during more than one stage of recovery (table 3). Given this situation, secondary and tertiary recovery may be defined as stages rather than methods of recovery. Thus, *secondary recovery* may refer to any improved recovery method used following primary production, and *tertiary recovery* to any improved recovery method used to recover additional oil after secondary recovery. In this manual, the terms *secondary recovery* and *tertiary recovery* will refer to stages of recovery rather than to methods.

The term *enhanced oil recovery* is used by the petroleum industry and the federal government in a variety of ways. It may be defined as all the methods used to introduce artificial drive into a reservoir in order to produce oil unrecoverable by primary recovery methods. It may refer to some or all of the methods that improve displacement and fluid movement in addition to providing reservoir pressure. Federal tax laws contain criteria that must be met before certain types of improved recovery projects can qualify for a reduction in the Windfall Profit Tax (see the Internal Revenue Code of 1954, as amended, Section 4993). Because of the lack of consensus on the definition of *enhanced oil recovery,* this manual will avoid using the term and will instead use the term *improved recovery* to refer to all methods of introducing artificial drive into a reservoir.

OTHER METHODS OF INCREASING PRODUCTION

During primary or improved recovery, production of oil and gas from a reservoir may be increased by the use of artificial lift and well stimulation methods.

Artificial Lift

About 90 percent of U.S. oil production employs some type of artificial lift. The most common methods are gas lift, sucker rod pumps, hydraulic pumps, and submersible pumps.

Gas lift. If a supply of gas is economically available and the amount of recoverable oil justifies the expense, gas lift is commonly

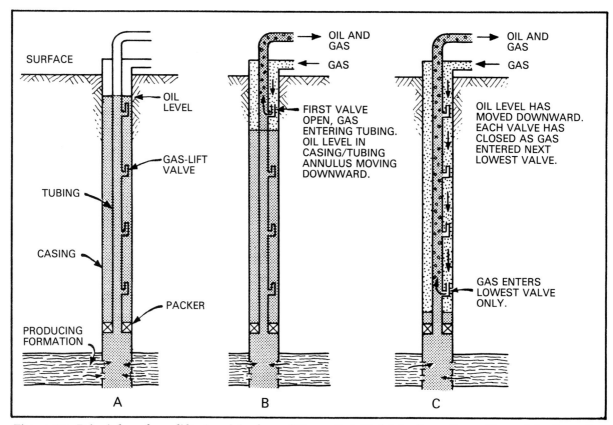

Figure 21. Principles of gas lift. *A*, original conditions; *B*, initial injection of gas; *C*, gas injection after expansion has occurred.

used to increase production (fig. 21). In the gas-lift process, gas is injected into the fluid column of a well, where it expands and aerates the oil. The lightened oil exerts less pressure than the formation and is forced out of the wellbore by the higher formation pressure. Gas may be injected continuously or intermittently, depending on the producing characteristics of the well and the arrangement of the gas-lift equipment.

Sucker rod pumps. The use of sucker rod pumps is commonly known as rod pumping. The well fluid is lifted to the surface by a subsurface pump located at or near the bottom of the well. Up-and-down motion is imparted to the rod pump by a string of sucker rods connected to a surface prime mover. The weight of the rod string and the fluid is counterbalanced by weights attached to the surface reciprocating beam or beam pumping unit or by air pressure in a cylinder attached to the beam (fig. 22).

Hydraulic pumps. Hydraulic pumping is a method of artificial lift that may be used to pump several wells from a central source and that can lift oil from depths of more than 10,000 feet. Two reciprocating pumps are coupled and placed in the well. One pump functions as an engine that drives the other. The fluid used to drive the subsurface engine is oil from the well itself, drawn from a surface tank and pumped downhole by a standard engine-driven pump. If a single string of tubing is used, power oil is pumped down the tubing string to the pump, which is seated in the string; and a mixture of power oil and produced fluid is returned through the casing-tubing annulus.

28

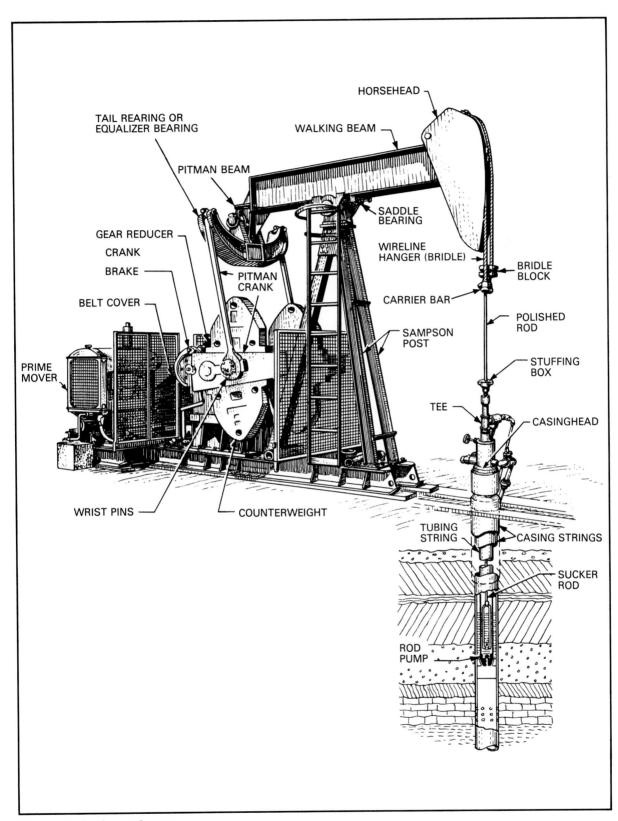

Figure 22. Sucker rod pump

If two parallel strings are used, one supplies the power oil to the pump while the other returns the exhaust and the produced oil to the surface.

Submersible pumps. In many older fields, wells produce a large volume of water compared to the volume of oil. In order to be economically justifiable, the well must produce large volumes of fluid. In these circumstances, a submersible centrifugal pump may be installed below the level of fluid in the well, either in the tubing or in the casing. The pump is usually driven by an electric motor and consists of a series of rotating blades that impart centrifugal motion to lift the fluid to the surface. Since both the motor and the pump are submerged in well fluid, the electric current is supplied through a special heavy-duty armored cable strapped to the tubing.

Well Stimulation

Well stimulation refers to methods used to increase the permeability of the rock surrounding a production well. The methods most commonly used today are acidizing and hydraulic fracturing.

Acidizing. Originally, acidizing was used to increase production in older wells that had developed permeability problems. Today, two types of acidizing are regularly used when wells are completed so that production can begin at the highest possible rates. The first type is fracture or high-pressure acidizing. The second is matrix or low-pressure acidizing.

Fracture acidizing, also called acid fracturing, is used to increase the natural permeability of carbonate formations as well as to repair damage from drilling and completion operations. Acid is forced into the formation at a pressure high enough to cause the rock to crack. The combined action of acid and high-pressure fluid enlarges existing channels, creates new channels, and clears out substances blocking the rock pores. Acid fracs are performed as a standard procedure in many oil fields.

Matrix acidizing is used to remove materials clogging rock pores, such as drilling muds, completion fluids, mud cake, and the clay deposits common in many sandstone formations. Acid is sent downhole at a pressure below the fracture pressure of the formation. The acid may be allowed to soak for a few hours to remove mud cake, or pumps may be used at low pressures, pushing acid a short distance into the formation to dissolve substances blocking the passage of oil and to widen natural fissures and channels. Matrix acidizing can be used in formations with a nearby gas cap or water zone where fracturing could cause uncontrollable gas or water production.

Hydraulic fracturing. *Hydraulic fracturing* is the high-pressure injection of fluids such as water, oil, or acid to create penetrating reservoir fractures. The method is particularly useful with sandstone reservoirs.

A hydraulically formed fracture tends to come back together after the fluid pressure is reduced unless the fracture is propped open. Therefore, the injected fluid is also used to carry a propping agent or *proppant* such as sand; nutshells; or beads of aluminum, glass, or plastic. Spacer materials are used between the particles of the proppant to ensure its optimum distribution.

Self-Test

2: Production Energies in Oil Recovery

[Multiply each correct answer by four to arrive at your percentage of competency.]

Fill in the Blank and
Multiple Choice with One or More Answers

1. During primary recovery, the type of energy used to produce oil is _____

   _____.

2. Explain how water drive works. _____

   _____

3. Pressure decline is likely to be most rapid in reservoirs with natural _____

   _____ drive.

4. Give an example of a combination drive. _____

   _____

5. _____ Gas lift works by –
   A. displacing oil from the pores.
   B. pumping gas and oil to the surface.
   C. lowering the viscosity of oil in the pores.
   D. aerating the oil in the wellbore.
   E. accomplishing all of the above.

6. Submersible pumps are used when _____

   _____

   _____

   _____.

7. The purpose of well stimulation is _____

   _____

   _____.

8. List two methods of well stimulation.

   (1) _____

   (2) _____

9. The amount of gas compared to the amount of oil being produced is called the _____

_____.

10. Define *improved recovery*. _____

_____

_____

11. _____ Improved recovery may be used to—
    A. displace oil from reservoir rock.
    B. improve fluid-flow patterns.
    C. enlarge or create new rock pores.
    D. increase formation pressure.
    E. accomplish all of the above.

12. The four major types of improved recovery are:
    (1) _____
    (2) _____
    (3) _____
    (4) _____

13. The most commonly used improved recovery method is _____

_____.

14. _____ Immiscible gas injection is used to—
    A. mix with the oil and displace it.
    B. aerate the oil in the wellbore.
    C. maintain reservoir pressure.
    D. make thick oils flow more freely.
    E. accomplish all of the above.

15. Miscible gas injection, chemical flooding, and thermal recovery may be called advanced recovery methods because they not only restore formation pressure but they also

_____

_____

_____.

attractive force that holds together molecules of the same substance. *Adhesion,* on the other hand, is the tendency for molecules of one substance to cling to those of a different substance. Both of these forces are caused by electrical attraction and repulsion occurring in the atoms and molecules of substances.

Properties of Liquids

The combined action of cohesion and adhesion creates certain properties in liquids. These properties include surface tension, wetting, and capillarity.

Surface tension, the tendency of a liquid to maintain as small a surface as possible, is caused by the cohesive attraction between the molecules of a liquid. This property can be demonstrated by filling a glass with water and placing a needle on top of the water (fig. 24). If placed carefully, the needle will float, even though it is heavier than water. It is held on the surface of the water by a thin film of molecules. Below the surface, each water molecule is attracted in all directions by surrounding molecules; thus, the forces of attraction tend to balance one another. Water molecules at the surface,

however, are attracted downward and sideways, but not upward, creating an elastic surface so strong that the molecules do not separate under the weight of the needle. The demonstration with the glass of water is an example of surface tension at the interface between a liquid and a gas. Surface tension, known as *interfacial tension,* also exists at the interface between two liquids that do not mix, such as water and oil.

Wetting, a second property of liquids, is the adhesion of a liquid to the surface of a solid. A liquid is said to wet a solid when the attraction of the liquid molecules to the solid (adhesion) is greater than the attraction of the liquid molecules to each other (cohesion). The property of wetting can also be seen in the water-and-needle demonstration (fig. 24). The random movement of water molecules causes some of them to bounce against the walls of the glass. The adhesive attraction of water to glass causes these water molecules to cling to the glass. At the same time, the clinging molecules are attracted cohesively to other water molecules on the surface of the water. As a result of these combined forces, the edge of the surface area is pulled upward. Certain

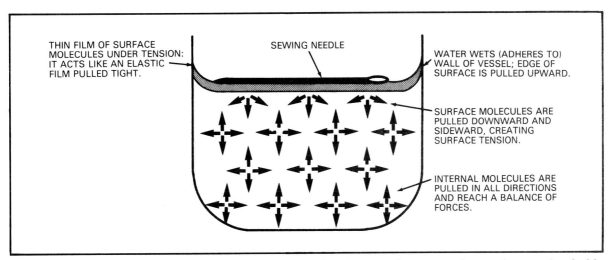

Figure 24. Demonstration of surface tension. The needle is heavier than water, but surface tension holds it up on the water's surface.

liquids such as mercury do not wet solids, because the cohesion of mercury molecules is stronger than their adhesion to solids.

The combined effects of surface tension and wetting bring about *capillarity,* the rise or fall of liquids in small-diameter tubes or tube-shaped spaces. Rising action can be observed by placing several glass tubes with different diameters in a bowl of water (fig. 25*A*). As water molecules adhere to the

sides of a tube and cohesion pulls other molecules up behind them, the water rises higher and higher in the tube. The smaller the diameter of the tube, the higher the water will rise above the level of the surrounding liquid. A falling rather than rising action occurs when the cohesion of a liquid exceeds its adhesion, as in the case of mercury (fig. 25*B*). When glass tubes are placed in a container of mercury, the liquid in the tubes separates from the sides of the tubes, drops below the level of surrounding liquid, and assumes a convex surface shape (arching upward). The convex surface is caused by *surface tension,* the tendency of a liquid to form as small a surface as possible by assuming a spherical shape.

Oil and Water Behavior in the Reservoir

In a reservoir, surface tension, wetting, and capillarity influence the way in which hydrocarbons and water migrate and the efficiency of oil displacement through waterflooding. These forces may work together to trap oil in pore spaces.

In reference to reservoirs, *wettability* is the relative adhesion of reservoir fluids to the grains of rock surrounding them. Rocks may be classed as oil-wet, water-wet, or intermediate on a continuum between these two extremes. The rocks in most oil reservoirs are intermediately water-wet to completely water-wet. In *water-wet* reservoirs containing both oil and water, water coats the rock grains and fills the smallest pores. Oil is held as droplets or islands in the middle of the larger pore spaces. Figure 26*A* represents conditions in a water-wet formation before, during, and after waterflooding. Early in the drive, a part of the reservoir is not yet swept by the waterflood. The rock grains are coated with water, and much of

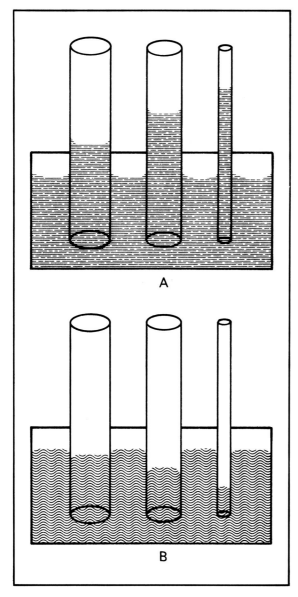

Figure 25. Capillarity. *A*, water-air interface; *B*, mercury-air interface.

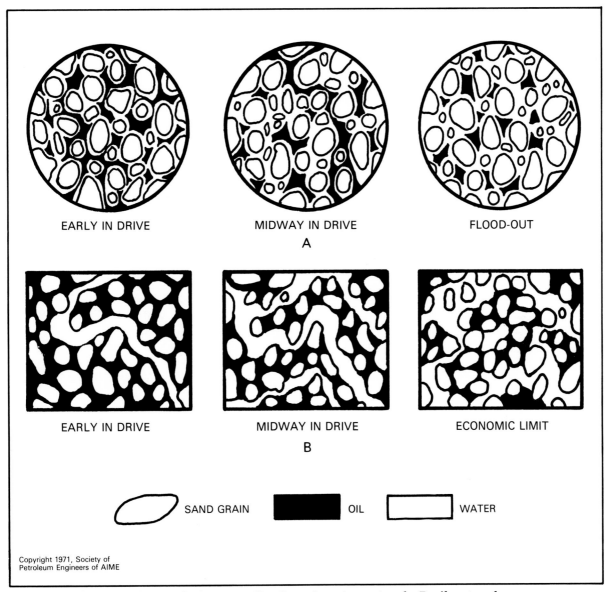

EARLY IN DRIVE   MIDWAY IN DRIVE   FLOOD-OUT

A

EARLY IN DRIVE   MIDWAY IN DRIVE   ECONOMIC LIMIT

B

SAND GRAIN   OIL   WATER

Figure 26. Fluid distribution during waterflooding. *A*, water-wet rock; *B*, oil-wet rock.

the remaining pore space contains oil. Later, channels develop from the flow of oil and water. Some channels reach dead ends, and some droplets of oil are passed by and trapped. Finally, at flood-out, only isolated islands of oil remain.

Rocks are called *oil-wet* if oil instead of water adheres to rock surfaces. In an oil-wet reservoir, oil coats the rock grains and fills the smaller pore spaces (fig. 26*B*). As water

enters the formation early in the drive, it forms a weaving channel. As the drive continues, the first channel enlarges and extends into side channels. As the drive approaches the economic limit, water is flowing freely through the main channel but is no longer displacing oil by enlarging side channels. Residual oil is left in the smaller pore spaces, and a film of oil coats rocks in the larger channels.

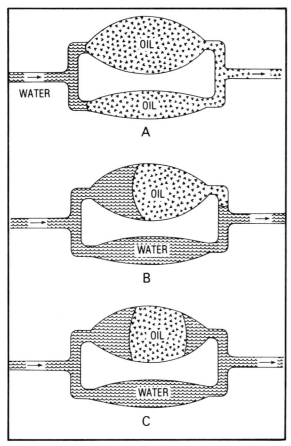

Figure 27. Pore model showing how capillary forces displace oil and trap residual oil. *A*, initial conditions; *B*, during displacement; *C*, after displacement. (Reprinted with permission of *Petroleum Engineer International*)

Capillarity both helps and hinders the displacement of oil, as shown in figure 27, a conceptual diagram of two pore spaces with a common inlet and outlet. In *A*, water from the flood begins to enter both pores. Because the pore walls are water-wet, capillarity helps draw water forward through the upper and lower pores, displacing oil. In *B*, water has displaced all the oil from the smaller pore but only part of the oil from the larger pore. Water moves faster through the smaller pore because the smaller diameter increases capillary pressure and because the amount of oil to be moved is much less than in the larger

pore. For these reasons, water moving through the smaller pore reaches the common outlet before all the oil in the larger pore has been displaced. At this point, a capillary counterpressure develops at the narrow "back door" of the larger pore. Pressure in this narrow outlet is greater than the pressure in the larger part of the same pore, at the upstream face of the oil droplet. The capillary counterpressure at the outlet prevents the remaining oil from moving out of the larger pore. *C* shows an oil droplet trapped after the two capillary pressures have reached a balance of forces. Because injected water has already established a flow path through the smaller pore, continued waterflooding will not displace the trapped oil.

Determining Displacement Efficiency

During waterflooding, of course, engineers cannot actually focus on a microscopic section of pore space. However, they can get information about displacement efficiency by analyzing core samples taken from the reservoir. Operating personnel play an important role in obtaining these cores, which require careful handling. Samples taken before waterflooding can be tested in a laboratory to yield information on initial and residual oil saturations, part of the data needed to predict how much oil will be recovered by waterflooding.

SWEEP EFFICIENCY

Sweep efficiency is the percentage of total reservoir volume that can be contacted by injected water. *Areal sweep* refers to the horizontal area that can be swept, and *vertical sweep* refers to the cross-sectional area that can be invaded.

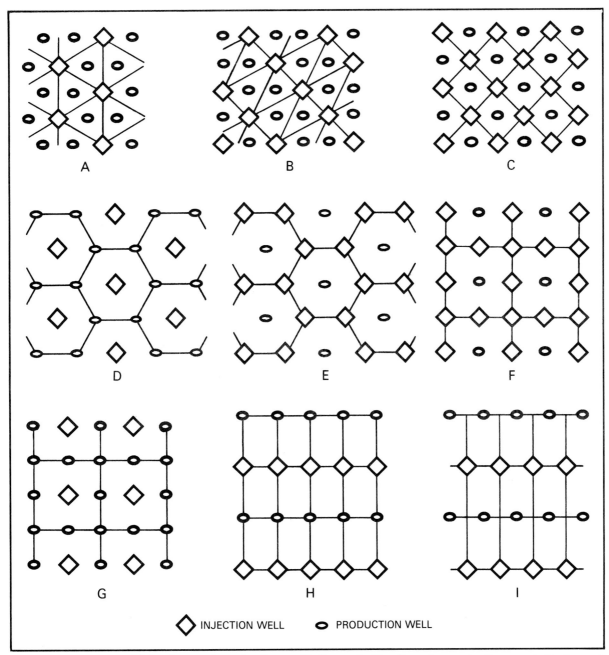

Figure 28. Patterns of injection and production wells for waterflooding. *A*, regular four-spot; *B*, skewed four-spot; *C*, five-spot; *D*, seven-spot; *E*, inverted seven-spot; *F*, normal nine-spot; *G*, inverted nine-spot; *H*, direct-line drive; *I*, staggered-line drive.

Areal Sweep Efficiency

Engineers generally arrange the injection and production wells of a waterflood project according to standard *injection patterns* (fig. 28). The five-spot pattern is most common; it consists of a square with wells at each corner and in the center. The repetition of these squares creates an interlocking pattern in which four injection wells surround each production well and four production wells surround each injection well.

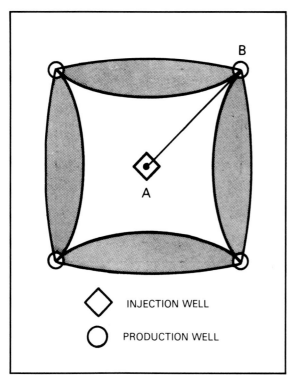

Figure 29. Areal sweep efficiency in a five-spot pattern

A direct line between an injection well and a production well (fig. 29, line *AB*) is the shortest distance that injected water must travel to break through at a production well. Water moving along this line will reach the production well before water that travels any other path. By the time the injected water that moves along the straight line arrives at the production well, only a portion of the area between the two wells will have been invaded. The horizontal area that has been swept, divided by the total area, equals the areal sweep efficiency at a certain point in time or at a certain water cut, expressed as a percentage.

Sweep efficiency is strongly affected by the *mobility ratio,* the ratio of the mobility of the driving fluid (in this case, water) to that of the driven fluid, oil. *Mobility,* which is determined by relative fluid viscosity and rock permeability, is the ease with which a

fluid flows through a reservoir. A mobility ratio of 1 indicates that injected water and oil flow through the reservoir with equal ease. A ratio of 10 means that water flows ten times more easily than oil.

Engineering studies have developed charts that show the specific effect of mobility ratio on sweep efficiency (fig. 30). As mobility ratios increase beyond 1, reservoir conditions are likely to become increasingly unfavorable for waterflooding. Water moves through or around the oil, seeking the path of least resistance toward a production well, arriving sooner, and sweeping less area between wells than water at lower ratios. Mobility ratios below or near 1 yield higher sweep efficiency and are considered favorable for waterflooding.

The completeness of sweep-out also depends on how long water injection can be continued. Production usually becomes uneconomic when the *water cut* – the percentage of water in the produced

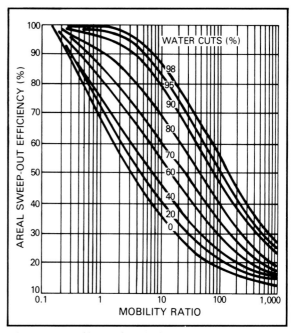

Figure 30. Effect of mobility ratio on the areal sweep-out efficiency of the five-spot pattern at various water cuts. (Reprinted with the permission of *Petroleum Engineer International*)

42

fluid—moves into the range of 90 to 98 percent. For example, if the mobility ratio in a five-spot pattern is 1 and the economic limit of the water cut is 95 percent, about 98 percent of the pattern area will be swept by the time the well has to be abandoned (fig. 30). If the mobility ratio rises to 10, areal sweep efficiency will decline to 85 percent.

Vertical Sweep Efficiency

Cross sections (such as fig. 14, p. 15) reveal many nonuniformities in the structure of petroleum reservoirs. *Reservoir heterogeneities* may include lenticular formations, pinch-outs, faults that cut across the reservoir, shale barriers, changes in porosity, and—probably most important—variations in permeability. These heterogeneities can have an important effect on sweep efficiency. Successful waterflooding is greatly aided by the even advance of water through the pores. The presence of heterogeneities decreases sweep efficiency by causing water to progress irregularly and permitting capillary forces to trap bypassed oil. Water enters the more permeable regions first and flows through them faster. Less permeable regions are entered later and less completely. The front of the waterflood, following the path of least resistance, reaches the production wells before some of the lower-permeability regions have been completely flushed. Oil in those regions is left behind (fig. 31).

The force of gravity also contributes to inefficient waterflooding. Because water is heavier than oil, injected water tends to move along the bottom of a reservoir. This *gravity segregation* of oil and water results in incomplete sweep of the upper regions.

Vertical sweep efficiency, the percentage of a reservoir's cross-sectional area contacted by injected water, is a measure of the

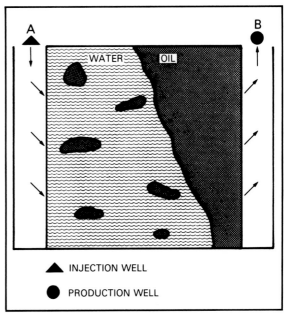

Figure 31. Cross section of a formation during a waterflood. As the water advances, it bypasses certain portions of the reservoir. Vertical sweep efficiency measures the percentage of a reservoir's cross-sectional area contacted by the waterflood.

flood's effectiveness in overcoming heterogeneities in the vertical dimension. Vertical sweep efficiencies typically range from 75 to 90 percent.

WATERFLOOD PERFORMANCE

Predicting Performance

When an oil company considers a waterflooding project, engineers try to predict how the reservoir will respond. Their most basic calculation, the amount of additional oil that can be recovered, indicates the potential benefit of the project; against this recovery, company decision makers must weigh the cost of the project. Engineers can also forecast details of project performance such as the rate at which the reservoir can accept injected water. This type of information is useful in planning and operating the project.

The prediction of recovery from the field as a whole can be done with a hand calculator. The more detailed prediction of performance requires the use of a large computer and a sophisticated program. A sample set of calculations is shown in figure 32; it may be followed by a reader who has had a beginning course in algebra. The choice of a simplified or a more detailed prediction depends on the apparent potential of a reservoir for waterflooding and the amount of detail needed to make decisions

---

1. ESTIMATED INITIAL OIL SATURATION

   A. **Assumptions and Reservoir Data**

   The hypothetical reservoir has produced 12 percent of the original oil in place by solution-gas drive.

   Laboratory tests show the following oil reservoir volume factor (a factor that corrects for compressibility and gas dissolved in reservoir oil, calculated by dividing the volume at reservoir conditions by the volume at standard conditions in an oil stock tank on the surface):

   1.30 at original saturation pressure
   1.21 at current saturation pressure

   Connate water saturation (percentage of pore volume) equals 8 percent.

   B. **Formula for Initial Oil in Place**

   $$S_{oi} = \frac{1 - S_{wc}}{B_{oi}}$$

   where

   S_{oi} = initial saturation of oil in place
   B_{oi} = oil formation volume factor at original pressure
   S_{wc} = connate water saturation

   C. **Calculation with Substituted Reservoir Data**

   $$S_{oi} = \frac{1 - 0.08}{1.30} = \frac{0.92}{1.30} = 0.708 \text{ STBO/RVB}$$

   Initially, each barrel of reservoir pore volume held 0.708 stock tank barrels of oil (STBO).

2. PREDICTED RESIDUAL OIL AFTER WATERFLOODING (BARRELS)

   A. **Assumptions and Reservoir Data**

   Production will continue until the water cut reaches 99 percent, corresponding to an average water saturation of 70 percent of the reservoir pore volume.

   The formation volume factors are the same as in step 1.

   B. **Formula for Residual Oil per Barrel of Pore Volume in the Swept Part of the Reservoir**

   $$S_o = \frac{1 - S_w}{B_o}$$

   where

   S_o = residual oil per barrel of pore volume in the swept part of the reservoir
   B_o = oil formation volume factor at current saturation pressure
   S_w = average water saturation

Figure 32. Sample calculation for oil recovery by waterflooding

C. **Calculation with Substituted Data**

$$S_o \frac{1 - 0.70}{1.21} = \frac{0.30}{1.21} = 0.250 \text{ STBO/RVB}$$

At the end of waterflooding, each barrel of swept reservoir pore volume will retain 0.250 stock tank barrels of oil.

3. PREDICTED RECOVERY FROM WATERFLOODING

A. **Assumptions and Reservoir Data**

Step 3 of the calculation employs mostly simple arithmetic and the results of steps 1 and 2.

B. **Total Oil Recovery**

Original oil in place per barrel of pore volume (PV)	0.708 STBO/RVB
Minus residual oil per barrel of PV after flooding	-0.250 STBO/RVB
Equals total oil recovery per barrel of PV	0.458 STBO/RVB

C. **Total Recovery as a Percentage of Original Oil in Place (OOIP)**

Total recovery per barrel of PV	$100 \times \dfrac{0.458}{0.708} = 64.7 \text{ percent}$
Divided by OOIP per barrel of PV	

D. **Percentage of Recovery by Waterflooding**

Total recovery percentage	64.7 percent of OOIP
Minus primary recovery	-12.0 percent of OOIP
Equals additional recovery by waterflooding	52.7 percent of contacted OOIP

E. **Correction for Heterogeneities**

A correction formula (not shown here) accounts for reservoir nonuniformities by factoring in (1) the variation of permeability distribution and (2) the mobility ratio.

F. **Corrected Prediction of Waterflood Recovery**

The corrected estimate of recovery by waterflooding is 49 percent of the original oil in place.

4. PREDICTED RESIDUAL OIL AFTER WATERFLOODING (PERCENTAGE OF OOIP)

A. **Assumptions and Reservoir Data**

Step 4 uses data from step 3 and the assumption that the reservoir has produced 12 percent of the OOIP by solution-gas drive.

B. **Residual Oil Available for Tertiary Recovery**

OOIP	100 percent
Minus oil produced by solution-gas drive	- 12 percent
Minus oil produced by waterflooding	- 49 percent
Equals residual oil saturation	39 percent

Figure 32 – Continued

about the project. The simpler method may be used to make a preliminary judgment about reservoir potential, and further study may be indicated by the results.

A perfect method of prediction, according to one of the foremost waterflooding authorities, would account fully for three kinds of effects:

1. fluid-flow effects resulting from permeability, wettability, pore size, connate water saturation, and other conditions;

2. well-pattern effects, including the effect of mobility ratios on sweep efficiency; and

3. reservoir heterogeneity effects such as variations in permeability and barriers to flow.

A computer prediction of reservoir behavior during waterflooding attempts to come close to such perfection. It can include all the major variables and can agree closely with actual results in simple systems. A simulation can predict the performance of individual wells over a period of many years, including injection and producing rates and expected water-oil ratios.

There are, however, some practical limitations on the applicability of computer simulations. They rest on a large number of simplifying assumptions that limit their validity. Moreover, computer models require very detailed descriptions of reservoirs and their fluids. This information may be unavailable or may take considerable time and money to gather.

Managing the Waterflood

The best pressure at which to begin waterflooding is the original bubble point. Because viscosity is lowest at this pressure, the mobility ratio and areal sweep are improved. For economic reasons, waterflooding may not be started until the pressure drops below the original bubble point; the purchase of water injection equipment earlier in a field's life delays the point at which production will become profitable.

Engineers monitor waterflooding as it proceeds in order to confirm or correct their calculations and to make necessary adjustments in project operation. The effects of unforeseen reservoir heterogeneities can be reduced in certain ways. Injection can be shifted to a different interval within the reservoir, and the location of injection wells can be changed. Sometimes, new *infill wells* are drilled in the space between existing production wells in order to recover isolated pockets of residual oil.

Despite these adjustments, a reservoir's performance may differ from the prediction. One possible explanation is the *resaturation effect.* As the waterflood advances, it may move oil toward rocks of lower permeability, where gas space has developed during primary production. Some oil may be forced into these rocks and then bypassed by the waterflood as it seeks the path of least resistance.

Range of Applications

Waterflooding has broad applicability to a wide range of reservoir and operating situations. When used for pressure maintenance during primary production, waterflooding can be applied to deep reservoirs.

Engineers generally believe that water-wet reservoirs flood better than those with oil-wet rock. Mobility ratio and sweep efficiency are also more favorable, generally, in water-wet reservoirs. Nevertheless, many oil-wet reservoirs have been flooded successfully.

Advantages and Drawbacks

The advantages of waterflooding are that it is relatively inexpensive and predictable to use. Under ideal circumstances, waterflooding can be expected to recover 50 to 60 percent of the original oil in place. The remaining oil will consist of about 20 to 40 percent residual oil in the water-swept region and a remainder of bypassed oil in the resaturated, unswept regions of the reservoir. If certain factors affect performance adversely, however, as much as 70 percent of the oil may remain after approximately 10 percent primary recovery and 20 percent recovery from waterflooding.

A major limitation of waterflooding is that even under the most favorable circumstances, it cannot displace all of the oil from a formation. First, water is immiscible with oil. Consequently, when water and oil come into contact, the resulting interfacial tension and capillarity create a resistance to fluid flow. Water tends to channel through pores of higher permeability, bypassing oil accumulations in lower-permeability areas. The trapped oil cannot be displaced by additional injection of water because of the strength of capillary forces. Second, water is denser than oil and tends to travel through the lower regions of a reservoir. As a result, sweep-out of the upper regions is incomplete.

Self-Test

3: Waterflooding

(Multiply each correct answer by four to arrive at your percentage of competency.)

**Fill in the Blank and
Multiple Choice with One or More Answers**

1. Define *waterflooding*. _____
   _____
   _____

2. A disadvantage of waterflooding is that _____
   _____
   _____
   _____.

3. The adhesion of a liquid to the surface of a solid is called _____.

4. Interfacial tension is the surface tension between _____
   _____.

5. When water molecules bounce up against a glass container, they cling to the glass because of the force of _____.

6. If a needle is carefully placed on the surface of water in a container, the surface molecules will not separate under the needle's weight because of the force of _____.

7. The smaller the diameter of a passageway through rock, the _____ (stronger/weaker) will be the force of capillarity.

8. Reservoirs in which oil coats the rock grains are called _____.

9. Capillarity hinders the displacement of oil by _____
   _____
   _____
   _____.

10. The success of a waterflood project depends on two factors of efficiency: (1) _____ and (2) _____.

48

11. Define the following terms:

(1) water cut _____

_____

_____

(2) bubble point _____

_____

_____

(3) injection well _____

_____

_____

(4) infill drilling _____

_____

_____

(5) areal sweep _____

_____

_____

12. The injection pattern being examined by the man on the front cover of this lesson is an example of a(n) _____ pattern.

13. The percentage of a reservoir's total volume that can be contacted by water during a waterflood is called the _____.

14. If water has a mobility more than ten times that of the oil in a certain reservoir, a waterflood of the reservoir is likely to be unfavorable because _____

_____

_____

_____.

15. If a reservoir has about the same permeability throughout, what will be the effect on a waterflood of this reservoir? _____

_____

_____

16. _____ If oil in a certain reservoir is much lighter than water, the most likely result during a waterflood of the reservoir will be —

A. partial sweep of the upper regions.
B. an even advance of the waterflood.
C. increased displacement efficiency.
D. movement of the oil downward.
E. all of the above.

17. Two factors taken into account in predicting oil recovery from a waterflood are —

(1) _____

(2) _____

18. _____ A disadvantage of computer simulations of reservoir performance is that —

A. they can predict only short-term results.
B. they can predict only for the reservoir as a whole, not for individual wells.
C. they require detailed, costly data about the reservoir.
D. they rest on simplifying assumptions that limit their validity.
E. they have all the above limitations.

4

Chemical Flooding

OBJECTIVES

Upon completion of this section, the student will be able to:

1. Define *polymer* and list three properties of polymers that are useful for improved recovery.

2. List the uses and the disadvantages of polymer solutions in improved recovery.

3. Define *micelle* and list the components of a micellar solution.

4. Identify the properties of a microemulsion and describe the process by which a microemulsion displaces oil from rock pores.

5. Describe the steps of a micellar-polymer flood.

6. Identify the reservoir conditions required for successful use of a micellar-polymer flood.

7. List the advantages and the disadvantages of micellar-polymer flooding.

8. Define *alkaline flooding* and identify four mechanisms by which alkaline solutions are believed to displace oil from rock pores.

9. Explain why alkaline flooding is a little-used improved recovery method.

10. Compare micellar-polymer flooding with other improved recovery methods in terms of production cost and recovery efficiency.

INTRODUCTION

Even under the most favorable conditions, drive fluids such as water and gas leave a large residue of oil—usually 25 to 50 percent—in the reservoir. One important reason is that these fluids are not miscible (do not mix) with oil, and the resulting interfacial and capillary forces resist displacement of the oil from rock pores. Water and gas can also have densities and mobilities different from oil, so they tend to sweep only certain regions of the reservoir (gravity segregation) and to finger past the oil (unfavorable mobility ratio). Reservoir heterogeneities present another problem; they can cause channeling through high permeability areas. Chemical flooding is an attempt to overcome inefficiencies of displacement or sweep by adding chemicals to water injected into a reservoir. One type of chemical flooding, polymer flooding, is used to make the viscosity and mobility of the drive fluid compatible with that of the oil and to improve sweep. Two other types, micellar and alkaline flooding, are used to create a drive fluid miscible with oil.

POLYMER FLOODING

Polymer flooding has been used as an adjunct to waterflooding and to chemical flooding with detergentlike surfactants. In a few instances, it has also been used as an improved recovery method in its own right. The two main functions of polymer flooding are the control of drive-water mobility and the control of fluid-flow patterns in reservoirs.

Nature of Polymers

Polymers are long, chainlike, high-weight molecules formed by the linkage of thousands of repeating chemical units

called monomers. To illustrate the size of these molecules, if the diameter of a molecule's links were increased to ¼ inch, the chain would extend over ¼ mile.

Polymers have three properties of importance for improved recovery: the ability to increase water viscosity, the ability to decrease effective formation permeability, and the ability to change polymer viscosity with rate of flow.

Small amounts of polymers dissolved in drive water increase the viscosity of the water, slowing the progress of the water through a reservoir and making it less likely to finger around oil in lower-permeability rock. Polymer-thickened water can also be used as a buffer between injection fluids of different viscosities to prevent them from mixing together during chemical flooding.

Certain polymers can decrease the effective permeability of reservoir rock by adsorption and entrapment. In *adsorption*, a layer of polymer molecules coats the walls of pores, reducing the size of the pores. *Entrapment* occurs when a polymer molecule enters a pore but is unable to leave because the exit opening is too small. The flow of water is reduced, but oil is still able to pass through. This capability of reducing permeability can be used during primary production to decrease the amount of water in produced fluids by blocking off reservoir layers producing water and allowing oil to enter the wellbore. The same capability can be used during improved recovery projects to turn water away from established channels toward lower-permeability areas that might otherwise be bypassed. Thus, polymer-thickened water can help to make the effective permeability of a reservoir more homogeneous and to make the sweep of water toward production wells more even. The viscosity must be adjusted to the pore sizes of a particular reservoir in order to prevent polymer molecules from clogging pore openings. An extremely heavy polymer would not be applied to rock with very small pores but would be appropriate in a reservoir with large fractures or other channels that are allowing drive water to move too rapidly toward production wells. When very heavy polymers are required to plug large channels, polymers are combined with dissolved metal ions to form thick gels. The gels must be carefully directed to the correct areas in order to prevent the blockage of oil flow.

Certain polymers are *pseudoplastic*—they decrease in viscosity and increase in mobility with increased flow rate. The viscosity increases again when the flow rate slows down. This property makes it possible to pump the polymers into injection wells at economical rates. As the polymers flow through rock and their flow rate decreases, they thicken and form a barrier to additional injected drive water.

Two polymers, polyacrylamides and polysaccharides, are used in improved recovery. *Polyacrylamides* are produced synthetically by combining carbon, hydrogen, oxygen, and nitrogen to form the basic unit, an amide monomer. *Polysaccharide biopolymers* (also known as Xanthan gums) are produced by a microorganism grown commercially by fermentation. They come from the remains of a shell grown by the microorganism.

The two types of polymers have different properties that limit their use to specific situations. Polyacrylamides, the most commonly used polymers, can be used to lower effective reservoir permeability as well as to increase the viscosity of drive water. They produce the highest viscosity when dissolved in fresh water. Polyacrylamides may lose viscosity when dissolved in water with

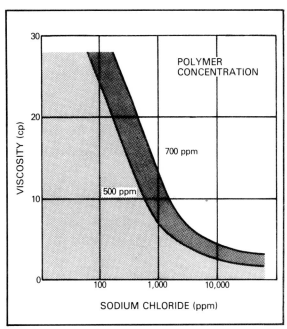

Figure 33. Effect of salinity on the viscosity of a polyacrylamide solution

a high salt concentration (fig. 33) or with multivalent *ions*—atoms or groups of atoms charged either positively or negatively as a result of losing or gaining electrons. (*Valence* refers to the combining power of a chemical element, measured by the number of bonds an atom of the element can form with hydrogen.) A preflush of the reservoir before polymer flooding may ease this problem by lowering reservoir concentrations of salt and minerals, such as calcium and magnesium, that release ions. In addition, polyacrylamide molecules can suffer *shear degradation*—breakdown into smaller units when subjected to rapid changes in flow rate and direction. When the force applied to a liquid causes one layer of the liquid to move more quickly than another layer, a resistance or stress is built up by the sliding of the two layers past each other. This resistance is known as *shear stress*. The faster the change in flow rate, the greater is the shear stress. This stress causes polymers

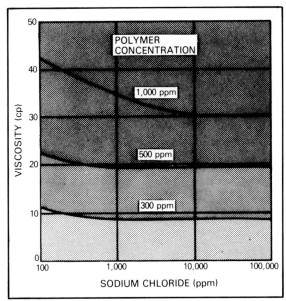

Figure 34. Effect of salinity on the viscosity of a polysaccharide biopolymer solution in deionized water

to break down and lose their useful properties. At very high rates of flow, polyacrylamides lose their pseudoplastic nature and become more rather than less viscous. They may clog pores at the injection well so that injection fluid cannot pass. Polyacrylamides must be handled carefully during mixing and injection to minimize shear degradation, and unavoidable breakdown may have to be offset by increasing the initial concentration of the polymer.

Polysaccharide biopolymers can be used in low-permeability reservoirs because they do not reduce the permeability of reservoir rock. They improve sweep efficiency solely by increasing the viscosity of water. They are not altered by changes in flow rate and direction and are easier to handle during injection operations. Polysaccharide biopolymers are not as sensitive to salt as polyacrylamides (fig. 34) and can be used effectively in brine, but hard water can cause the formation of gels. Polysaccharide biopolymers are highly susceptible to

breakdown by bacterial action, and must be protected by the application of a biocide.

Range of Applications

Polymer flooding can be used in a variety of situations to control the mobility of drive water, to combat channeling, and to alter the flow patterns of reservoirs. As a displacement fluid, its efficiency is no greater than that of water, and its sole value is the improvement of sweep efficiency.

Polymers have been used to improve waterflood performance by decreasing the effective permeability of the formation to water, and thus decreasing the volume of produced water. Permeability to oil remains about the same, however, allowing oil to flow into the wellbore. Polymers also improve waterflood performance by decreasing water mobility. An effective tenfold reduction of water mobility requires a 0.02 to 0.05 percent polymer solution in an injection the size of 20 to 40 percent pore volume. Even when a waterflood is expected to give good recovery without polymers, they may be added to help overcome unforeseen reservoir heterogeneities and to ensure that the waterflood will be successful.

Polymer flooding is used most widely as a mobility buffer during chemical flooding with surfactants. A quantity of injected polymer-thickened water forms a zone of viscosity transition between the initial injection of surfactant solution and the final injection of drive water.

As noted earlier, the use of polymers may be limited by reservoir permeability and by concentrations of salt and ions in reservoir fluids. Too little or too much salinity can rule out the use of polymers; an upper limit for salinity is about 150,000 parts per million (ppm).

Advantages and Drawbacks

A small concentration of polymer—250 to 1,500 ppm—can change the viscosity of injected water, reducing its mobility and increasing sweep efficiency. Polyacrylamides can also decrease the effective permeability of reservoir rock, thus decreasing water mobility further and creating a more even sweep pattern. Polysaccharide biopolymers can be used in brine water and can tolerate rapid changes in flow, making them easier to handle during mixing and injection operations. In addition, polymers are nontoxic and noncorrosive.

The chief disadvantage of polymers is the tendency of polymer molecules to break down under the influence of temperature, chemicals or bacteria in reservoir fluids, or shear stress. Reservoir temperatures should not exceed 250°F. Polyacrylamides must be mixed with fresh, oxygen-free water to prevent contaminants from reducing the viscosity, and they must be handled by surface equipment that minimizes shear degradation. Polysaccharide biopolymers must be filtered and treated with a biocide to prevent breakdown from bacterial action. A second disadvantage is that polymers do not improve displacement efficiency, one of the two major factors influencing the success of improved recovery.

Economic and Technical Outlook

Polymer flooding is expected to have limited future application in comparison with other improved recovery methods. An authoritative study found polymer flooding to be the method of choice in only 7 out of 245 reservoirs screened. By contrast, surfactant flooding was chosen for 58 of the reservoirs.

As an aid to waterflooding, polymer flooding is expected to produce an incremental recovery of something under 500 million barrels, or less than 1 percent of total potential recovery by all improved recovery methods. Although its cost is lower than that of some of the other improved recovery methods, the recovery efficiency is also lower—about 4 percent of the residual oil in place. The most important application of polymer flooding will probably continue to be its use as a mobility buffer in surfactant flooding. The combination of surfactant and polymer flooding provides effective improvement of both displacement and sweep efficiency.

MICELLAR-POLYMER FLOODING

Micellar-polymer flooding is a two-step improved recovery method in which a surfactant-water solution is injected into a reservoir to free oil from rock pores, and polymer-thickened water is then injected to drive the oil toward production wells.

Nature of Surfactants

Television commercials advise viewers to "wash out those dirty oil stains" with detergents. Petroleum engineers have taken this advice by using detergentlike chemicals to mobilize residual oil trapped in reservoir rock. Soaps and synthetic detergents contain a basic ingredient called a *surfactant,* or surface-active agent. A surfactant molecule consists of a polar (electrically charged) group of atoms linked to a long nonpolar chain of hydrocarbon units. The polar end is soluble in water, and the nonpolar chain is soluble in oil. The dual attraction of surfactant molecules to water and oil gives them the capability of reducing surface tension and breaking up the oil into small droplets that can be moved from rock pores by drive water.

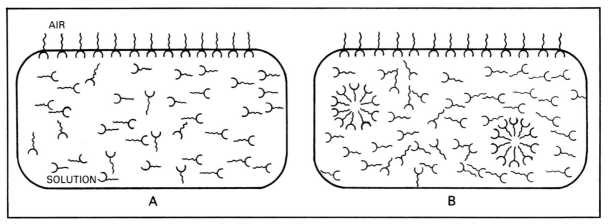

Figure 35. Properties of surfactants. *A*, dilute solution in water; *B*, micellar solution in water and oil.

When small amounts of surfactants are dissolved in water, they are dispersed randomly in the water because molecular forces pull the molecules in all directions (fig. 35*A*). However, when introduced into a water-oil system, the surfactant molecules begin to migrate toward the interface between the two substances. The water-liking end of each surfactant molecule seeks water and the oil-liking end seeks oil. From this tendency to concentrate at the interface, the molecules get the name *surface-active agent.* As one end of a surfactant molecule dissolves in oil and the other in water, a bridge is formed between the oil and water,

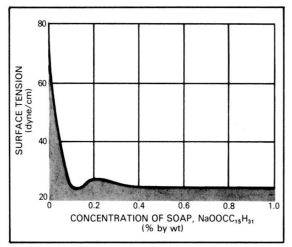

Figure 36. Effect of surfactants on the surface tension of water

lowering the interfacial tension between them (fig. 36).

When the amount of surfactant is increased to a critical concentration, the surfactant molecules begin to surround oil droplets, forming round clusters called *micelles* (fig. 35*B*). The hydrocarbon ends of the surfactant molecules dissolve in the drop of oil captured inside the micelle, and the polar ends extend outward and dissolve in the surrounding water. The dissolved polar ends are negatively charged, and they repel each other. As a result, the micelles push each other away and the oil droplets are prevented from coming back together into a separate zone of oil. This type of oil-water mixture is called an *emulsion,* because it is a system consisting of a liquid dispersed in an immiscible liquid. It is an *oil-in-water* emulsion because the oil is distributed in the water (fig. 37*A*). In some mixtures, surfactants may form a *water-in-oil* emulsion. Then, the surfactant molecules form micelles around drops of water, and the water is dispersed in the oil (fig. 37*B*). Both types of emulsions are effective in displacing oil during chemical flooding.

An emulsion formed by surfactants is a special type known as a *microemulsion,* because it has certain properties that are

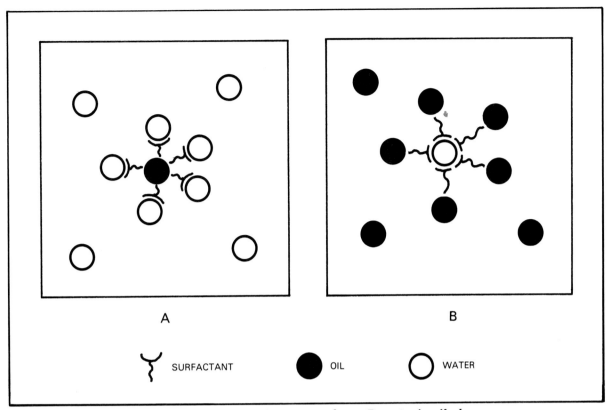

Figure 37. Phases of microemulsions. *A*, oil-in-water phase; *B*, water-in-oil phase.

like emulsions and other properties that are more like solutions. In emulsions, the dispersed liquid is the size of microscopically small droplets, so emulsions have a cloudy appearance. The substances mixed in emulsions tend to separate and settle out after a period of time, like the oil and vinegar in a salad dressing. One substance is dispersed in the other. In an emulsion of oil and water, for example, the water is either dispersed in the oil, or the oil in the water. *Solutions* have different qualities because they are mixtures of two miscible substances in which the dissolved substances subdivide to molecular size. Salt water is an example. Solutions are optically clear. They are stable mixtures; the substances mixed together do not separate. The substances are completely mixed together.

Like emulsions, microemulsions are mixtures of two immiscible substances, and they often have oil-in-water or water-in-oil phases. However, under certain conditions they can also have a phase in which oil and water are not organized in spherical micelles but are lined up in alternating layers. Like solutions, microemulsions are optically clear and stable. These special properties are due to the presence and activity of the surfactant.

Displacement Mechanism

Micellar microemulsions are used during chemical flooding to break up oil into small droplets. Then the pressure of the injected fluid moves the emulsified oil out of the rock pores. The way a micellar emulsion operates in a reservoir can be visualized by using the pore model introduced in section

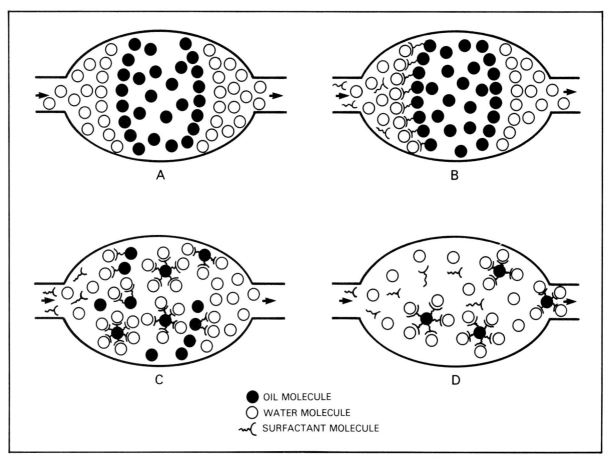

Figure 38. Surfactant-flooding displacement mechanisms in an ideal pore model. *A*, oil droplet trapped in pore; *B*, surfactant reduction of surface tension and capillary action; *C*, penetration of injection fluid into oil droplet and formation of micellar solution; *D*, removal of oil-water mixture from pore.

3, which showed the entrapment of an oil droplet by capillary forces. Figure 38 illustrates the mechanism by which surfactant flooding overcomes this condition. Before the arrival of the micellar surfactant solution, the oil and the water are separate phases, and surface tension at the interface between the two phases is high (fig. 38*A*). When the surfactant molecules reach the oil-water interface, oil-liking parts of the surfactants begin to attach themselves to oil molecules, and water-liking groups connect with water molecules (*B*). Now, the surface molecules of the two phases are no longer repelled by unlike molecules but are attached to similar molecules. As a result,

surface tension falls. When it drops below a certain point, the pressure of the injected fluid overcomes the force of surface tension and begins to move the trapped oil. After surfactant molecules have penetrated the interface between the two liquids, separate phases of oil and water no longer exist (*C*). Micelles begin to form, usually of oil droplets dispersed in water. The surfactants help keep the water and oil mixed, and the pressure of injected fluid now moves the microemulsion out of the upper pore and into the main outlet (*D*).

As the microemulsion of surfactant, oil, and water moves on through the reservoir, it mixes with more oil and water, and

58

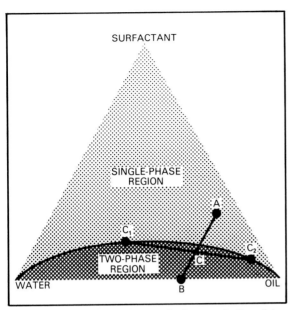

Figure 39. Composition and phase relationships of a mixture of surfactant, oil, and water

changes occur in its characteristics. These changes can be visualized on a ternary, or three-sided diagram (fig. 39).

Each corner of the diagram represents a pure component of the three-component system. The arc from bottom left to bottom right of the diagram separates the single-phase from the two-phase region. In the single-phase region, the three components form a micellar microemulsion. Oil and water are made miscible by the action of the surfactant and coexist in a single phase. The important feature of this microemulsion for oil recovery is that the leading edge will mix with the reservoir oil that it contacts, displace it miscibly, and recover essentially all of it. Most of the oil will be pushed into a bank ahead of the microemulsion and driven toward a production well.

As the surfactant solution continues to mix with more reservoir fluids, the microemulsion changes composition and eventually deteriorates into two or more phases. The micellar solution (point A) mixes with reservoir oil and water (point B).

A mixture of about one-third oil and two-thirds water falls along the line from A to B. When the composition reaches point C in the two-phase region, the microemulsion will separate into two or more phases. (For the purpose of illustration, only two phases are shown.) The mixture will be predominantly water at C_1, and predominantly oil at C_2. In the two-phase region, the microemulsion will no longer exhibit the miscible characteristics that allowed it to emulsify oil. However, oil will continue to be mobilized because of the lowering of interfacial tension by the surfactant solution, and an immiscible type of displacement will continue. The stages of these types of displacement are shown in figure 40.

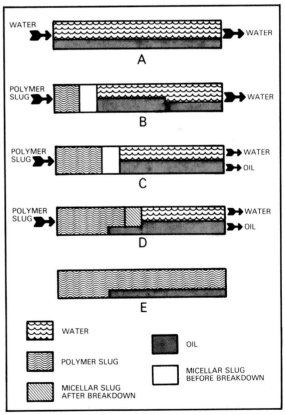

Figure 40. Changes in type of displacement during micellar-polymer flooding. A, initial state of core; B, and C, miscible displacement; D, immiscible displacement; E, final state of core.

The easiest way to ensure good displacement efficiency during surfactant flooding would be to inject surfactant continuously until breakthrough occurs at the production wells, comparable to the continuous injection of water during waterflooding. However, the high cost of the chemicals limits the amount that can be used. A measured quantity, or *slug*, of surfactant solution is injected; and the flood is designed to prevent loss of the surfactant due to adsorption or chemical reactions with substances in reservoir fluids, to slow down dilution of the surfactant by reservoir and drive fluids, and to enhance immiscible displacement after the slug separates into two or more phases. Miscible displacement is prolonged by preceding the surfactant injection with a water preflush, adjusting the composition of the surfactant slug to be compatible with reservoir conditions, and following the slug with an injection of polymer-thickened water. Immiscible displacement is enhanced by keeping the interfacial tension between the surfactant solution and the residual reservoir oil as low as possible.

Process Description

Before a micellar-polymer flood is begun, engineers analyze data about reservoir conditions to determine whether a flood is feasible, whether a preflush will be necessary, and what kind of surfactant solution will be best for that particular reservoir. Data are collected from core samples and reservoir fluids.

A micellar flooding operation proceeds in four injection steps (fig. 41).

1. *Preflush injection.* An injection of water may be needed to improve the compatibility of reservoir fluids with the surfactant slug.

2. *Surfactant-solution injection.* A solution of surfactant, cosurfactant, electrolyte, oil, and water is injected to achieve displacement of residual oil.

3. *Polymer-solution injection.* A solution of water and polymer (thickened water) is injected to push the surfactant slug through the formation. The viscosity of the solution is decreased gradually so that it approaches that of the water which will follow.

4. *Water injection.* Water is used to drive the polymer slug and the surfactant slug until the project reaches its economic limit.

Preflush. A preflush may be required in order to adjust chemical conditions in the reservoir. It can also be used to provide information on reservoir flow patterns by employing tracers.

Four reservoir conditions that affect chemical flooding are salinity or salt concentration, temperature, ion concentration, and rock material that adsorbs surfactant. Too low or high salinity in reservoir fluids can change the phase makeup of the surfactant solution and reduce its effectiveness. As noted earlier, high salinity can also cause loss of polymer viscosity. A preflush can be used to lower the reservoir's salt concentration to a more favorable level; and polysaccharide biopolymers, which are less sensitive to salinity, can be selected. High temperatures are another cause of surfactant breakdown. The presence of magnesium and calcium ions in reservoir fluids can cause surfactant molecules to separate and settle out of solution, with a resulting separation of the microemulsion into oil and water phases. Ion concentration can be lowered by a preflush of the reservoir. Certain types of clays are known to adsorb surfactants, thus weakening the surfactant

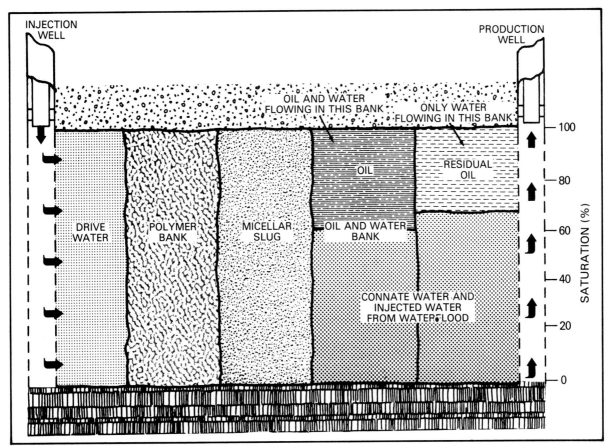

Figure 41. Stages of micellar flooding

solution and causing breakdown of the miscible displacement process. If a chemical such as sodium silicate is added to the preflush, it is adsorbed by the rock instead of the surfactant, lowering surfactant loss.

Surfactant solution. In addition to water, surfactant solutions have four constitutents: surfactants, cosurfactants, electrolytes, and oil. The surfactant usually chosen is petroleum sulfonate, because it is less expensive than other surfactants. Cosurfactants are added to adjust the viscosity of the surfactant solution, help keep the solution stable, and reduce adsorption of the surfactant onto reservoir rock. Economic considerations usually restrict the choice of cosurfactants to light alcohols. More than one cosurfactant may be used.

An electrolyte, a salt such as sodium chloride, is also added to aid in the adjustment of viscosity. Oil is added to help form the micelles of the surfactant solution. Crude oil produced earlier from the field is generally chosen because of its low cost, ready availability, and compatibility with reservoir fluids.

The size of the surfactant slug is from 5 to 10 percent of reservoir pore volume. Five percent is the minimum needed to achieve efficient displacement, and 10 percent is probably the upper limit due to chemical costs.

The composition of surfactant solutions is designed to be compatible with the reservoir environment based on the results of core analyses and other data on expected

61

reservoir conditions (table 4). A well-designed solution is compatible with reservoir salinity, has built-in safeguards against surfactant loss by adsorption or chemical reaction, and provides favorable interfacial tension for both miscible and immiscible flooding.

TABLE 4
Range of Compositions for Micellar Solutions

Constituent	Percentage of Micellar Slug
Surfactant	4 to 10
Cosurfactant	about 4
Electrolyte	about 1
Hydrocarbon	4 to 80
Water	10 to 92

SOURCE: E. F. Herbeck, R. C. Heintz, and J. R. Hastings, *Fundamentals of Tertiary Oil Recovery*. Dallas: Harcourt, Brace, Jovanovich, Energy Publications Division, 1977.

Polymer solution. If water alone were used to drive the surfactant slug, the contrast in viscosities would cause fingering, break up the slug, and reduce sweep efficiency. The polymer solution acts as a buffer; it protects the surfactant slug from the drive water injected as the last step in micellar-polymer flooding. In order to be large enough to protect the surfactant slug, the buffer should equal at least 50 percent of reservoir pore volume. Mobility at the buffer's leading edge should be equal to or less than that of the micellar solution. The viscosity of the buffer is adjusted by the addition of polymers.

Contrasts in mobility should also be minimized between the buffer and the drive water that follows it. This aim can be achieved by gradually changing the concentration of polymer in the mobility buffer (fig. 42). Grading produces a more favorable mobility ratio between the drive water and the mobility buffer and reduces the amount of chemical needed.

Water. The final step is to inject water to drive the micellar slug and the mobility buffer toward breakthrough at the production wells. This phase of operations is the same as for waterflooding.

Range of Applications

Generally speaking, fields that have responded well to waterflooding are good candidates for recovery by micellar-polymer flooding. Such fields are likely to have good permeability and sweep efficiency. However, a successfully waterflooded reservoir may not respond well to chemical flooding if the residual oil saturation is too low or if the reservoir is too heterogeneous. Residual saturation should be in the range of 25 to 30 percent of pore volume. Sweep problems encountered during waterflooding will have an even greater impact on the outcome of chemical flooding, because high-cost chemicals, unlike water, are injected in limited quantities. In addition to residual oil saturation, the conditions affecting the success of surfactant flooding are as follows:

1. Surfactant flooding is applicable only to sandstone reservoirs, not to carbonate reservoirs, because they adsorb excessive amounts of surfactants.

2. The salinity of reservoir fluids must be favorable for the particular composition of the micellar solution being injected. The optimal salinity must be determined for each reservoir. Salinities over 150,000 ppm percent are generally too high to permit effective surfactant flooding.

3. Magnesium and calcium ion concentrations in reservoir fluids must be low enough that they do not cause breakdown of the surfactant molecules. According to one estimate, the

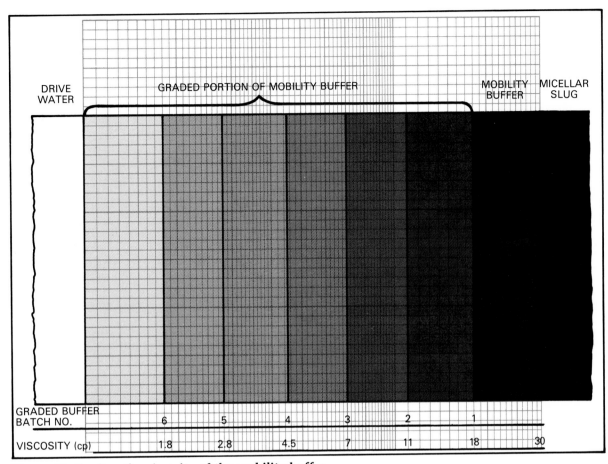

Figure 42. Grading the viscosity of the mobility buffer

concentration of ions should be less than 2,000 ppm for good results. Future development may produce surfactant solutions that are less sensitive to multivalent-ion exchange.

4. A permeability of at least 20 millidarcys (md) is required, and the variation in permeability should not be too great. Permeability determines injection rates and affects flood patterns. Great contrasts in permeability (for example, a variation ratio of over 7 to 1) will cause uneven distribution of the surfactant solution. High-permeability zones will receive too much chemical, low-permeability zones will receive too little, and the overall recovery efficiency will suffer (fig. 43).

5. Reservoir heterogeneity must be moderate. Chemical flooding does not work well in reservoirs with extensive faulting, lenticular formations surrounded by impermeable shale, fluid movement primarily through fractures, or bottom-water drive. The reservoir must permit horizontal flow of fluids. By the mid-1980s, micellar-polymer flooding will probably be applicable to formations with minor heterogeneities if good well-to-well correlations are available to assist engineers in planning the flood.

6. An oil viscosity of about 20 centipoise (cp) is considered the limit for the application of micellar-polymer flooding. *Centipoise* is a measure of

63

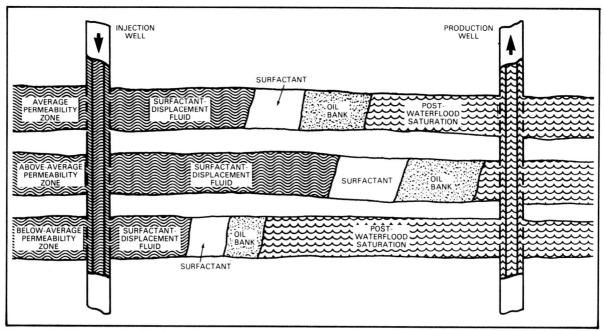

Figure 43. Uneven distribution of chemicals caused by differences in reservoir permeability

liquid resistance to flow. Oil viscosities higher than 20 cp would require increasing the viscosity of the surfactant and polymer slugs to achieve a favorable mobility ratio; the higher chemical concentrations required would cost too much.

7. Reservoir temperatures suitable for micellar-polymer flooding are ranging up to 200°F during the mid-1980s; technological developments are expected to increase this limit to 250°F later in the century.

Advantages and Drawbacks

Micellar-polymer flooding allows engineers to try to adjust the two factors of efficiency that account for improved recovery: displacement efficiency and sweep efficiency. In laboratory tests, displacement is essentially 100 percent efficient—that is, the surfactant solution displaces all the oil it contacts as long as the slug remains intact. In applications to actual reservoirs, the method has recovered about 50 percent of the oil left after waterflooding. Economic factors limit the size of the slug that can be injected, and the relatively small volume cannot remain intact throughout the reservoir. In addition, geological conditions prevent the surfactant slug from contacting all the oil present in the reservoir.

Operation of a micellar-polymer flood is essentially similar to waterflooding except for the addition of mixing facilities. Considerable use can be made of existing wells for production and, with some modifications, for injection. Indeed, the economics of the process are so marginal that use of existing surface equipment is almost always required. When micellar-polymer flooding is used as a tertiary recovery method in a mature field, total recovery will probably not be large enough to permit extensive redrilling and replacement of surface equipment.

One of the greatest hindrances to wider application of the method is the difficulty of

TABLE 5
Comparison of Recovery Potential, Efficiency, and Cost of Improved Recovery Methods

Process	Potential Recovery (Billion Barrels)[a]	Percentage of Total Improved Recovery[b]	Recovery Efficiency (Percentage of Residual Oil in Place)	Cost per Barrel[c] (Mid-1980 Dollars)
Micellar-polymer (surfactant)	5 to 12	28 to 23	30 to 43	35 to 46
Polymer (including alkaline)	0.3 to 0.9	2 to 2	4	22 to 46
Carbon dioxide	7 to 21	39 to 40	30 to 38[d]	26 to 39
Steam drive	4 to 17	22 to 32	25 to 64	21 to 35
In-situ combustion	1.6 to 2.0	9 to 4	28 to 39	25 to 36
Total	18 to 53			

SOURCE: Harry R. Johnson, "Prospects for EOR in the U.S.," *Well Servicing,* Jan.–Feb. 1982.

[a]Smaller estimates assume the state of development of methods in the early 1980s, and higher estimates assume an evolved technology that permits increased sweep efficiency. Increased efficiency is generally obtained by using blocking or mobility-control agents that direct oil-recovery injections to proper locations in the reservoir.
[b]Percentages add to more or less than 100 because of rounding.
[c]Cost data include profit.
[d]The figures in the source are 15 to 19 percent of original oil in place. These figures are equivalent to about 30 to 38 percent of residual oil in place.

predicting performance reliably. Because a great number of factors can affect actual performance, laboratory tests and field performance vary considerably, and the relation of models to actual conditions requires further refinement.

As noted earlier, the high cost of chemicals places significant limits on the application of the method. The investment in chemicals must be made at the very beginning of a project; however, if the surfactant flood follows waterflooding (the most common application), at least several months and up to two years may elapse before the wells produce anything but water. The combination of low predictability, high cost, and a long wait for an economic return has made micellar-polymer flooding appear a risky proposition to oil field operators during the early 1980s.

Economic and Technical Outlook

Micellar-polymer flooding has been called the Cadillac of improved recovery methods (table 5). Because of the high cost, no micellar-polymer project had proved commercially successful up to the beginning of the 1980s. At that time, operator ratings of profitability indicated that steam flooding was the most developed method and micellar-polymer flooding was the least developed (table 6). Polymer flooding took

TABLE 6
Operator Performance Evaluations
of Improved Recovery Methods
(to 1980)

TECHNIQUE	PERFORMANCE EVALUATION (Number of Projects)		
	Successful, Profitable	Promising	Discouraging
Steam	65	20	6
Polymer	10	4	3
In situ combustion	7	1	3
Carbon dioxide	2	5	1
Micellar-polymer	0	5	3

SOURCE: Harry R. Johnson, "Prospects for EOR in the U.S.," *Well Servicing,* Jan.–Feb. 1982.

second place for profitability, despite its limited recovery efficiency. The low ranking of micellar-polymer flooding may be explained partially by the government tax incentives in force at the time. The same tax break was received regardless of the recovery method used. Thus, if an operator chose a less expensive method, the tax incentive covered a larger portion of his risk.

Although micellar-polymer flooding is not expected to become commercially profitable until the mid to late 1980s (depending on a number of factors such as the price of oil and the rate of inflation), the method continues to receive a great deal of attention because of its potential for further development. Experts hope to develop less expensive surfactants and to find ways of applying the method more widely. Also, manufacturing facilities for surfactants and polymers are expected to grow, and with increased supplies, relative cost should decline.

ALKALINE (CAUSTIC) FLOODING

Alkaline or *caustic flooding* is a method of improved recovery in which alkaline chemicals such as sodium hydroxide are injected with a waterflood or a polymer flood. The chemicals react with the natural organic acids present in some crude oils to form surfactants within the reservoir.

Displacement Mechanisms

The surfactant created within the reservoir seems to improve oil recovery by several mechanisms. Like injected micellar surfactants, it reduces interfacial tension between water and oil, and it breaks up and emulsifies oil so that it can be moved out of the pores by water or a polymer solution. Under certain conditions, the alkaline solu-

tion may also change wettability. For example, in oil-wet reservoirs, oil tends to cling to pore walls, spreading out across the surface and filling tiny depressions and cavities (fig. 44A). The alkaline solution is thought to make the pore walls water-wet, causing the oil to draw away from the walls and to form droplets that can be swept out of the pores by the drive fluid (fig. 44B,C). This mechanism is useful in reservoirs where oil is still flowing. In water-wet reservoirs

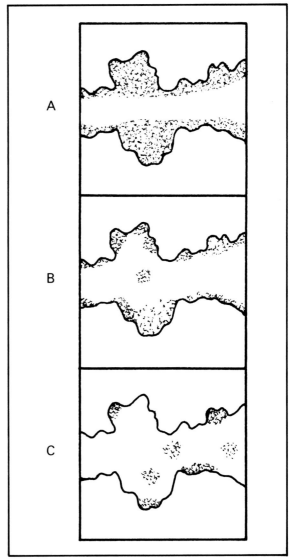

Figure 44. Oil displacement by alteration of rock wettability. *A,* oil-wet pore; *B,* beginning of wettability change; *C,* water-wet pore.

where residual oil is trapped in scattered locations, an alkaline solution may be used to change the rock to oil-wet, to promote water-in-oil emulsions, and thus to mobilize the remaining oil. A third mechanism appears to be entrapment. High-viscosity oil is emulsified by the alkaline solution and moves into new pores, where it is trapped by small pore openings. The trapped oil forces injection fluid into new areas of the reservoir, improving sweep efficiency. Finally, the alkaline solution is believed to break up and dissolve tough films that form at the interface between oil and water, allowing oil to flow more freely.

Process Description

Alkaline flooding has often been considered an improved type of waterflooding. It can be carried out by adding caustic chemicals to a waterflood or by mixing the caustic with a polymer solution. The polymer improves the mobility ratio and sweep efficiency. The process is similar to micellar-polymer flooding. Thorough laboratory testing and careful evaluation are necessary to determine reservoir suitability.

Range of Applications

Reservoir crude oil must have enough organic acid to be reactive with the caustic flood. The effectiveness of the method increases with acid content, which is generally highest in low-gravity oils. *API gravity*, a measure of oil density (fig. 45), correlates directly with crude oil acidity – the heavier the oil, the higher is its acid content. The National Petroleum Council has found the

---

- API gravity, expressed in degrees, measures the density or weight of an oil. API gravity is derived from specific gravity by the following formula:

$$\text{API gravity} = \frac{141.5}{\text{specific gravity}} - 131.5$$

- Specific gravity is the comparison of the density of one substance to the density of a standard substance such as pure water.

- According to the formula, a specific gravity of 1 is equal to an API gravity of 10°.

$$141.5 \div 1 = 141.5 - 131.5 = 10° \text{ API.}$$

- The higher the API gravity number, the lighter the oil, and the lower the API gravity number, the heavier the oil.

Specific Gravity	API Gravity	Comments
1.000	10°	A heavy oil.
0.985	12°	A heavy oil.
0.940	19°	Oils below 20° API make the best candidates for alkaline flooding and are also likely candidates for thermal recovery.
0.880	29°	The lightest oil (highest API gravity) considered suitable for alkaline flooding.
0.825	40°	Oils ranging from around 30° API to somewhat above 40° API are referred to as medium-gravity oils. They are considered most applicable to micellar-polymer flooding, carbon dioxide flooding, or hydrocarbon miscible injection.

Figure 45. API gravity and improved recovery processes

67

best candidates for alkaline flooding to be reservoirs in California with API gravities mostly below 20° and 30°. Although heavy oils with gravities below 20° API contain the acids necessary for alkaline flooding, they may also be recovered successfully by thermal methods, which are widely used in California.

Other application guidelines are similar to those for micellar-polymer flooding. Alkaline flooding is limited to sandstone reservoirs because carbonate rock reacts with the alkaline solution and neutralizes it. A low ion concentration is also necessary to avoid reactions that will neutralize the recovery chemicals, and the reservoir temperature must be below 200°F.

Advantages and Drawbacks

Chemicals for alkaline flooding are much less expensive than those used in surfactant solutions: surfactants alone cost over three times as much per pound as sodium hydroxide, the caustic most commonly used.

Alkaline flooding is expected to account for relatively little of the oil recovered by improved recovery processes – under 1 percent of the total. Table 5, p. 65 includes estimates for alkaline flooding along with those for polymer flooding. Alkaline flooding alone accounts for about half the combined amount, or fewer than 0.5 billion bbl.

Alkaline flooding seems to be the process of first choice for very few reservoirs. A National Petroleum Council screening found that only 4 out of 245 reservoirs were prime candidates for caustic flooding. In part, this small number is due to the need for high acid content in the crude oil and for other favorable reservoir conditions. In addition, thermal processes offer a preferable alternative for certain reservoirs.

Economic and Technical Outlook

Both alkaline flooding and steam drive can be used in reservoirs with high-viscosity oil. Alkaline flooding costs about the same as steam drive but has a much lower recovery efficiency (table 5). Some alkaline projects may prove commercially profitable because of the modest investment required, the effect of tax incentives, and the response of a particular reservoir. Overall, however, the best forecasts available in the early 1980s expected alkaline flooding to play a very small role in improved recovery compared to other processes.

Self-Test

4: Chemical Flooding

(Multiply each correct answer by three and add one bonus point to arrive

at your percentage of competency.)

Fill in the Blank and
Multiple Choice with One or More Answers

1. Define *polymer.* _____

   _____

   _____

2. Three properties of polymer solutions that make them useful for improved recovery are –

   (1) _____

   _____

   (2) _____

   _____

   (3) _____

   _____

3. _____ In improved recovery, polymer solutions are used to –
   A. lower the viscosity of oil.
   B. lower the mobility of drive water.
   C. alter the flow patterns of reservoirs.
   D. create a microemulsion with oil.
   E. accomplish all of the above.

4. A disadvantage of polymer solutions is that _____

   _____

   _____

   _____.

5. The five components of a surfactant solution are –

   (1) _____

   (2) _____

   (3) _____

   (4) _____

   (5) _____

69

6. _____ The properties of a microemulsion include –
   A. stability.
   B. an oil-in-water phase.
   C. a water-in-oil phase.
   D. optical clearness.
   E. all of the above.

7. A(n) _____ is added to a surfactant solution in order to adjust viscosity, keep the solution stable, and reduce surfactant adsorption.

8. The main purpose of a preflush is to _____

   _____

   _____.

9. Explain how a surfactant solution displaces oil from a rock pore. _____

   _____

   _____

   _____

   _____

   _____

   _____

   _____

   _____

   _____

   _____

10. After a slug of surfactant solution has been injected into the reservoir, the next step is to __

    _____

    _____.

11. The purpose of grading the viscosity of a mobility buffer is _____

    _____

    _____

    _____

    _____.

12. _____ Reservoir conditions required for successful use of a micellar-polymer flood include –

    A. moderate heterogeneity.
    B. low ion concentrations.
    C. high oil viscosity.
    D. at least 25 percent oil saturation.
    E. all of the above.

13. Micellar-polymer flooding gives engineers a chance to try to increase oil recovery by improving the (1) _____ and the

    (2) _____ of an improved recovery operation.

14. Three disadvantages of micellar-polymer flooding are –

    (1) _____

    (2) _____

    (3) _____

15. Define *alkaline flooding.* _____

_____

_____

_____

_____

16. _____ Alkaline solutions are believed to displace oil from rock pores by –

    A. increasing drive-water viscosity.
    B. improving reservoir flow patterns.
    C. changing rock wettability.
    D. emulsifying the oil.
    E. doing all of the above.

17. _____ API gravity (°API) measures oil –

    A. mobility.
    B. volume.
    C. density.
    D. temperature.

18. Give one reason why alkaline flooding is a little-used recovery method. _____

_____

_____

19. Of the improved recovery methods presently available, the method with the highest production cost is _____.

20. _____ Compared to thermal recovery methods such as in situ combustion and steam drive, the present recovery efficiency of micellar-polymer flooding is—

A. slightly higher.
B. much higher.
C. slightly lower.
D. much lower.

5

Carbon Dioxide Flooding

OBJECTIVES

Upon completion of this section, the student will be able to:

1. Compare the use of carbon dioxide with that of other injection gases in terms of cost and recovery efficiency.

2. Describe the process of multiple-contact miscibility by which carbon dioxide displaces reservoir oil.

3. Define *miscibility pressure* and list three factors that influence it.

4. Explain how carbon dioxide works to displace oil immiscibly.

5. Identify the factors that can cause poor sweep efficiency during a carbon dioxide flood.

6. Identify the reservoir conditions favoring and hindering successful use of carbon dioxide flooding.

7. List three reasons why the West Texas area is particularly suited for the use of carbon dioxide flooding.

INTRODUCTION

For many years, the oil industry has made a practice of returning produced natural gas into the reservoir in order to maintain reservoir pressure and to store the gas for future use. An added benefit of this practice has been the displacement of additional oil. Usually, displacement takes place immiscibly (by gas drive) because of the low injection pressures used and the composition of the gas. Other gases, such as nitrogen and flue gas, have also been used for pressure maintenance.

In the 1950s, the industry began to carry out gas-injection projects in search of a miscible process that would recover oil effectively during secondary and tertiary production. The injection gases used include the following:

1. liquefied petroleum gases (LPGs) such as propane and butane, followed by natural gas or natural gas and water;
2. methane enriched with other light hydrocarbons, followed by methane or methane and water;
3. methane under high pressure;
4. nitrogen under high pressure; and
5. carbon dioxide under pressure, alone or followed by water.

These gases are effective in displacing most of the oil in place in laboratory experiments, but their field use is limited by various economic considerations and practical problems of application. LPGs are miscible on first contact with crude oil and are applicable to many reservoirs. However, LPGs are now in such demand for marketing that their future role in improved recovery seems sharply limited. Methane, too, has a high market value, although enriched methane and methane under high pressure may be economical injection gases in areas without ready markets for natural gas. Methane and nitrogen are miscible with oil under high pressures, but their use is limited to deep reservoirs that can withstand the pressures without fracturing. Although displacement efficiency is high for all these gases, sweep efficiency is low because of the great differences in mobility and density between the gases and crude oil. The injected slug of gas has a strong tendency to finger past the oil and to break through prematurely into production wells, and gravity segregation is pronounced.

In the 1970s, the industry began to show a great deal of interest in carbon dioxide as a potential injection gas. Carbon dioxide shares the sweep efficiency problems of other injection gases, but to a lesser extent because of its greater density and its greater viscosity under pressure. Because carbon dioxide achieves miscible displacement at lower pressures than those required for methane and nitrogen, it is applicable to a greater number of reservoirs. And, in areas having a plentiful supply of carbon dioxide, it is an attractive choice because it is less expensive than methane and LPGs.

DISPLACEMENT MECHANISMS

Carbon dioxide can exist in a liquid, gaseous, or solid state. In the gaseous state, it is familiar as the bubbles in soft drinks; and in the solid form, it is known as dry ice. Liquid as well as gaseous carbon dioxide can be used to displace oil miscibly, but injected carbon dioxide is almost always in the gaseous state because its critical temperature (the temperature above which it exists only as a gas) is only 88°F, and most reservoir temperatures are higher. Gaseous carbon dioxide is dense and behaves like a liquid at pressures above 1,000 pounds per square inch absolute (psia). Under higher pressures, it is several times more viscous than methane under the same conditions.

Vaporization of Hydrocarbons

Unlike LPGs, carbon dioxide is immiscible with reservoir oil on first contact. However, under the right conditions of pressure, temperature, oil composition, and repeated contact, carbon dioxide can vaporize certain hydrocarbons from crude oil and extract them, forming a single-phase fluid that is miscible with crude oil. The miscible transition zone moves the crude oil toward the production wells. Because this process is dependent on certain conditions, including multiple contact between the carbon dioxide and the oil, it is called *multiple-contact, conditional,* or *dynamic miscibility.*

Vaporization consists of converting a substance from a solid or a liquid state into a gaseous state, or vapor phase. Crude oil is a mixture of hydrocarbons with a range of light to heavy molecular weights corresponding to the number of carbon atoms within the molecules. (*Molecular weight* is the sum of the weights of the atoms within a molecule.) Carbon dioxide can vaporize light hydrocarbons (molecules with 2 to 6 carbon atoms) and medium-range liquid hydrocarbons (7 to 30 carbon atoms), but it does not vaporize heavy solid hydrocarbons (over 30 carbon atoms). Therefore, even under ideal conditions, about 5 percent of reservoir crude oil will be left behind during carbon dioxide flooding. Vaporization and extraction of hydrocarbons by carbon dioxide works similarly to the process occurring when methane is injected under high pressure. However, unlike methane, carbon dioxide does not require the presence of light hydrocarbons in crude oil (2 to 6 carbon atoms) in order to form a miscible zone. Therefore, carbon dioxide can be used for improved recovery in wells that have been depleted of lighter hydrocarbons by gas drive during primary production.

Carbon dioxide will not vaporize hydrocarbons unless reservoir fluids are under high pressure. The *miscibility pressure,* at which vaporization begins to occur, is dependent on the temperature and density of the reservoir oil. Vaporization begins at pressures between 1,000 pounds per square inch (psi) and 2,000 psi at temperatures below 200°F with crude oils having gravities above 30° API. The denser the oil and the higher the temperature, the greater is the pressure required to attain miscibility. Additional pressure is required to reach the *optimum miscibility pressure,* at which enough hydrocarbons are vaporized to maintain a moving transition zone and to produce a profitable amount of oil. The maximum practical pressure is around 6,000 psi.

Immiscible Displacement Mechanisms

Carbon dioxide dissolves in oil, increasing the volume and decreasing the viscosity of the oil and allowing it to flow more easily. When reservoir pressure is sufficiently lowered by production, carbon dioxide dissolved in oil comes out of solution and displaces oil from rock pores. This process is identical to the solution-gas drive provided by natural gas originating in a reservoir.

PROCESS DESCRIPTION

Injection Process

Researchers have suggested a number of injection schemes for carbon dioxide flooding: continuous injection of carbon dioxide, injection of carbon dioxide followed by gas or water, alternating injection of carbon dioxide with gas or water, simultaneous injection of carbon dioxide with water, and combinations of carbon dioxide with solvents and heat injection. In some cases an initial injection of water may be needed to bring reservoir pressure to the level needed for miscibility. Reservoir simulation studies indicate that simultaneous or alternating injection of carbon dioxide and water is most effective in oil displacement. Continuous injection of carbon dioxide is thought to be most effective in reservoirs with steeply sloping pay zones and downward displacement of oil.

The most widely used injection scheme has been alternate injection of water and carbon dioxide (fig. 46). The initial slug of carbon dioxide equals about 5 percent of reservoir pore volume. Alternate injection of water and carbon dioxide continue until

the cumulative volume of injected carbon dioxide reaches a predetermined percentage of reservoir pore volume, typically 15 to 30 percent. Then water injection is continued until the conclusion of the project.

When carbon dioxide is injected into an oil-bearing formation at miscibility pressure, it first vaporizes the light hydrocarbons from the oil. As more carbon dioxide arrives at the front between the oil and the injection gas, the carbon dioxide vaporizes middle-range hydrocarbons, forming a transition zone between the two fluids. When enough hydrocarbons have been taken into this zone, it becomes miscible with the reservoir crude oil and pushes the oil in a bank toward the production wells. As the miscible zone moves forward, it disperses, and miscibility is lost. Newly arriving carbon dioxide vaporizes more

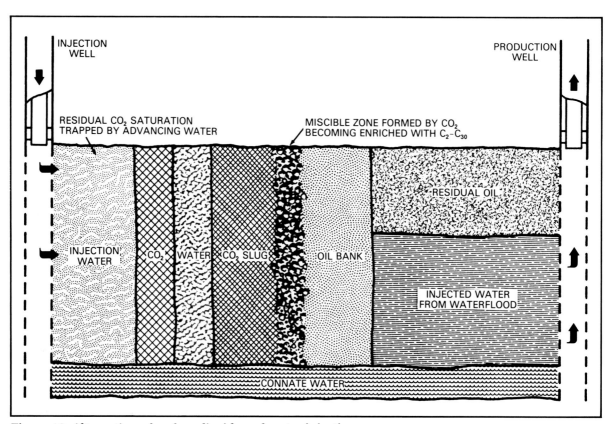

Figure 46. Alternation of carbon dioxide and water injection

hydrocarbons and regenerates the transition zone; this cycle of dispersion and regeneration continues along the path of displacement until the bank of oil is moved into the producing wellbores. Heavy hydrocarbons not vaporized by carbon dioxide or pushed along by the miscible zone are left behind as residue.

Design Problems

A major design problem in carbon dioxide flooding is the improvement of sweep efficiency. Carbon dioxide, with a viscosity typically 10 or 20 times less than that of oil, tends to take the path of least resistance through the oil, leaving large areas of the reservoir unswept. The mobility ratio of one project was around 5, as compared to the ratio of 1 necessary to achieve a sweep efficiency approaching 100 percent.

Under some reservoir conditions, the density of carbon dioxide approaches that of oil and water. However, when the density of carbon dioxide is lighter than that of other reservoir fluids, gravity segregation occurs and contributes to sweep inefficiency. The gas flows preferentially through the upper part of the reservoir, leaving a large portion uncontacted. The extent of gravity segregation depends on reservoir characteristics such as pay zone thickness and vertical permeability.

Reservoir heterogeneities also add to uneven sweep. Successful field tests have achieved sweep efficiencies around 50 percent. Although recovery from the swept portion is high, the size of the unswept area lowers overall recovery efficiency considerably.

The alternation of slugs of water with carbon dioxide is aimed at improving sweep efficiency by preventing fingering and gravity segregation. Injected water prevents the channeling of carbon dioxide by reducing the effective permeability of the formation to the gas. Water injection may cause as well as prevent sweep problems, however. In lower-permeability reservoirs, injected water may reduce the injection rate. Also, research indicates that in reservoirs with the complex porosity often found in carbonate rock, water may block the access of carbon dioxide to some pore spaces.

Some of the heaviest hydrocarbons are not vaporized by carbon dioxide and are left behind as asphaltene residues. These residues can also slow the injection rate by clogging small rock channels and interfering with the flow of carbon dioxide.

Another problem is predicting performance. Performance is studied by the use of physical models that simulate reservoir conditions and by computer models that quantify reservoir conditions and process variables. The results of these studies are used to design an effective flood for the particular reservoir. However, models are only as useful as the accuracy and comprehensiveness of the data used to create them. Physical models do not take reservoir heterogeneities into account. Computer models are also subject to error. Although the principal process variables have been identified, the models may not include all variables or may fail to assess accurately their relative importance under different conditions.

RANGE OF APPLICATIONS

Miscible carbon dioxide flooding is suitable for carbonate as well as sandstone formations and can be used in reservoirs lacking in natural gas and LPGs. It may not be feasible for use in reservoirs with large gas caps because large volumes of carbon dioxide may be required to establish miscibility pressure, and undesirable mixing may occur between the two gases. Like other injection gases, carbon dioxide is also unsuitable for reservoirs with large fractures that will channel the gas and cause premature breakthrough at production wells. The reservoir should have a minimum oil saturation of 25 percent.

Reservoirs must be deep enough to permit operation at miscibility pressure without fracturing. A safe miscibility pressure can be estimated by multiplying reservoir depth by 0.6 psi to obtain the fracturing pressure and subtracting a safety factor of 300 psi. For example, a formation with a depth of 3,000 feet would have a safe miscibility pressure of no more than 1,500 psi. This method is used only for screening purposes. Actual fracturing pressures must be determined by injection tests in the field. Some reservoirs, for example, may fracture at 0.45 psi or at 0.8 psi times reservoir depth.

As noted before, the main factors influencing the calculation of required miscibility pressure are the density of the crude oil and the reservoir temperature. The heavier an oil (the lower the API gravity number), the more pressure it requires. Heavier oils contain fewer of the vaporizable hydrocarbons in the light and middle ranges. Disregarding other factors, a 30° API oil requires 1,200 psi for beginning miscibility, and a 27° API oil requires 4,000 psi. Crude oil gravity must exceed 25° API, and a somewhat higher gravity—perhaps 28° API—makes operation of the flood easier. The miscibility pressure varies by 200 to 500 psi, taking the temperature into account. Pressure is also affected by the purity of the carbon dioxide. A small amount of methane in carbon dioxide will raise the required operating pressure. To reach the optimum miscibility pressure, several hundred additional pounds per square inch above beginning miscibility pressure will be needed.

ADVANTAGES AND DRAWBACKS

Carbon dioxide flooding appears to be the improved recovery method of choice for many reservoirs in which oil gravity is too high to favor thermal methods. It is also suited for carbonate reservoirs, which cannot be flooded by surfactants because the surfactants lose effectiveness when they react chemically with carbonate rock. The theoretical (laboratory) displacement efficiency of this method, as for all miscible recovery methods, approaches 100 percent. Compared to other injection gases, carbon dioxide achieves miscibility at lower pressures, making it applicable to shallower reservoirs. Being nonhazardous and nonexplosive, carbon dioxide is safer to work with than hydrocarbon injection gases.

On the minus side, carbon dioxide flooding has relatively poor sweepout efficiency compared to other improved recovery methods. Studies of West Texas waterflooded reservoirs indicate maximum oil recovery to be about 20 percent of the original oil in place (or about 30 to 40 percent of residual oil in place) using carbon dioxide flooding as the tertiary recovery method. These studies have been conducted with computer models calibrated on the results of actual field tests. The main limit on carbon dioxide flooding, however, is the availability of reliable, plentiful, inexpensive supplies of carbon dioxide. In locations such as West Texas, where such supplies are available, carbon dioxide flooding is profitable despite problems of sweep efficiency and low recovery efficiency.

Other problems include corrosion and deterioration of equipment. Carbon dioxide and water form carbonic acid, which is highly corrosive to equipment. Special metal alloys and coatings are used to combat corrosion. Coatings usually prove to be the more cost-effective method. Carbon dioxide also causes swelling of rubberlike materials called elastomers, which are used in downhole tools. Special resistant materials have been developed for packing components used in carbon dioxide flooding.

ECONOMIC AND TECHNICAL OUTLOOK

Gas Sources and Cost

Carbon dioxide can be found in reservoirs with natural gas or alone in nearly pure form. It can also be obtained as a by-product from power plants, chemical and fertilizer plants, and coal gasification plants. Naturally occurring carbon dioxide is the most likely source of sufficient quantities for use in improved recovery. Many of the known reserves are located in the states of Utah, Colorado, and New Mexico (fig. 47). This group of states is often referred to as the *Four Corners*—the other "corner" being formed by the state of Arizona.

Although carbon dioxide can be transported by refrigerated trucks or railroad tank cars, pipelines are the only economical means of transport from the source to the oil field. In order to justify the initial expense of constructing a pipeline, the carbon dioxide flooding project (or projects) must be sizable.

The cost of a barrel of oil obtained by carbon dioxide flooding includes expenditures for the purchase of carbon dioxide, transportation of the gas, tangible investment,

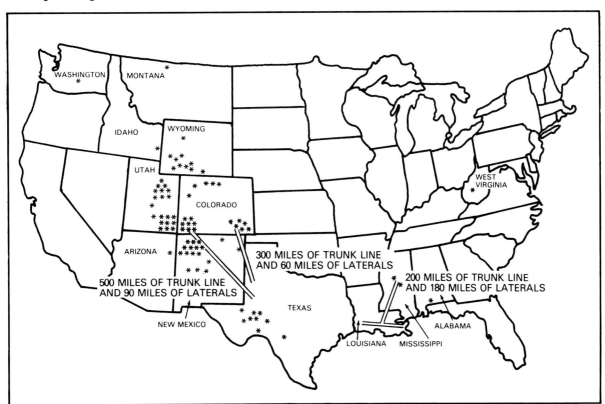

Figure 47. Major sources of naturally occurring carbon dioxide

fuel, carbon dioxide recycling, and well operation. Purchase and transportation of carbon dioxide account for 50 percent or more of the total production cost. *Tangible investment* includes expenditures for such items as compressors, separation equipment, well conversion, and the removal of hydrogen sulfide. Fuel is required not only to power compressors but also to supply heat in the carbon dioxide recycling process. As in other improved recovery operations, these costs must be paid through investments made at the beginning of a project; a period of time elapses before wells begin to produce and to bring a financial return. The range of per-barrel costs for carbon dioxide flooding is similar to that of other improved recovery methods – from $26 to $39 in mid-1980 dollars, according to estimates of the U.S. Department of Energy (table 5, p. 65). At favorable locations, the cost may run significantly lower than average.

Although production costs continue to rise, oil fields close to plentiful carbon dioxide supplies enjoy an economic advantage, since a large cost item is the construction and operation of pipelines to transport carbon dioxide (fig. 47).In the mid-1970s, for example, the highest cost for delivery of carbon dioxide to the injection well was about $2.50 per thousand cubic feet (Mcf), but the gas could be delivered to West Texas by pipeline from the Four Corners area for $0.65/Mcf (table 7). By the early 1980s, the cost of delivering carbon dioxide to West Texas had risen to $1 to $2/Mcf, but it was correspondingly higher when delivered to locations farther away. Naturally, areas such as the Permian Basin in West Texas that enjoy a favorable proximity to carbon dioxide supplies will be among the first to use carbon dioxide for large-scale improved recovery operations.

TABLE 7
Cost of Carbon Dioxide Delivered to
Various Geographical Areas
(mid-1970s)

Geographical Area	Cost (Dollars per Mcf)
Utah	0.50
New Mexico	0.65
Texas	
Districts 7C, 8, 8A, 9, 10	0.65
Districts 2, 3, 4, 5, 6	2.75
Louisiana	±1.00 if pipeline is built from Mississippi
	2.75 other pipeline
Mississippi	1.50
Colorado	1.50 for one-half of reservoirs
	2.75 for one-half of reservoirs
Oklahoma	1.50 for one-half of reservoirs
	2.75 for one-half of reservoirs
California	2.75
Others	2.75

SOURCE: National Petroleum Council, *Enhanced Oil Recovery: An Analysis of the Potential for Enhanced Oil Recovery from Known Fields in the U.S. – 1975 to 2000.* December 1975.

Gas Requirements and Supplies

In the 1970s, estimates of the amount of carbon dioxide required to recover a barrel of oil ranged from 2,000 to 3,000 Mcf. The actual amount of gas used varied greatly. Projects reported using from 5 to 26 Mcf per stock tank barrel of oil. Estimates of carbon dioxide requirements in the early 1980s ranged from 5 Mcf to over 10 Mcf/bbl. One reason for the variation is the use of different slug sizes, which generally vary from 15 to 30 percent of pore volume.

Although larger slug sizes increase incremental oil recovery, recovery efficiency decreases per unit of injected gas, and the cost per barrel of oil rises. In carbonate reservoirs, for example, the model used by the National Petroleum Council predicted that a 20 percent slug would bring about a 22 percent increment over primary and secondary recovery. Enlarging the slug to 40 percent of pore volume, the model predicted, would raise the increment to 33 percent. In other words, doubling the slug size of carbon dioxide would produce only a 50 percent increase in the recovery efficiency. The model predicted even poorer response in sandstone reservoirs. Increasing slug size from 25 to 35 percent would raise the recovery efficiency from 18.5 percent to only 20.5 percent.

In the mid-1980s, the total carbon dioxide requirement for the United States is expected to exceed presently known supplies of the gas (table 8). Some experts predict that insufficient reserves will limit the use of carbon dioxide flooding in improved recovery. Others expect that vigorous exploration will discover enough additional supplies to meet the nation's needs. Whatever the case, available supplies will most likely be used first in areas where costs are lowest, initial investment risk is the least, and the margin for profitable recovery is the greatest. Given the gap between requirements and supplies, it will be necessary to recycle carbon dioxide efficiently. Pilot tests have shown that when

initial costs are moderate and carbon dioxide is recycled, this improved recovery method is economically feasible.

Recovery Potential

Despite problems with sweep efficiency, the total oil recovery from carbon dioxide flooding may exceed that of any other improved recovery method – from 7 to 20 billion barrels, or about 40 percent of all improved recovery. One explanation for the high total is the wide applicability of carbon dioxide flooding. In a study of 245 reservoirs, the National Petroleum Council found 109 of them likely candidates for this method. Of the 109, 26 were candidates for both carbon dioxide flooding and other methods, and were assigned to other methods. Fifteen candidates were considered good prospects for carbon dioxide flooding, 24 were considered fair prospects, and 44 were considered poor prospects. Sixty-two of the reservoirs, including all 15 ranked as good prospects, were located in Texas.

Because of special geologic and economic conditions, 70 to 75 percent of all oil recovered by carbon dioxide flooding will

TABLE 8
Estimated Requirements for Carbon Dioxide, United States and Texas
(mid-1980s)

Area	Estimated Recovery by CO_2 Flooding (billion bbl)	Percentage of Total	CO_2 Requirement (trillion ft³)	Expected Supplies of Fresh CO_2 (trillion ft³)
United States	7 (low estimate)	NA	35	25
	20 (high estimate)	NA	200	32 to 50
Texas	5 (low estimate)	70	25	25
	15 (high estimate)	75	150	32 to 50

SOURCES:
E. L. Claridge, "CO₂ Takes the Spotlight in Texas," *Drill Bit*, March 1982, pp. 88, 90.
"Improved Economics and Technical Advances Point to a New Era of Miscible Flooding," *Journal of Petroleum Technology*, February 1982, p. 316.
Harry R. Johnson, "Prospects for EOR in the U.S.," *Well Servicing*, January–February 1982, pp. 51–55.

come from Texas reservoirs. Texas plays a large part in improved recovery because it contains about one-third of all known oil in the United States. Moreover, a higher proportion of the state's known oil is located in carbonate reservoirs than for the nation as a whole—46 percent of Texas oil compared to 28 percent of U.S. oil—and carbon dioxide flooding is especially suitable for many carbonate reservoirs. Primary and secondary recovery from carbonate reservoirs has averaged about 30 percent, leaving a large amount of oil as a target for tertiary recovery by carbon dioxide flooding.

In the Permian Basin area, several factors particularly favor carbon dioxide flooding. Reservoir temperatures are lower for a given depth than in other areas, permitting lower miscibility pressures. Because of the presence of large reservoirs in a single general area, pipeline construction is feasible; most of the 2,000 miles of planned carbon dioxide pipeline will go into West Texas. And, as already noted, the area's proximity to natural sources of carbon dioxide makes production costs significantly lower than for most other sites.

Technical Advances

Researchers are currently seeking information and techniques that will lead to increased recovery by carbon dioxide flooding. Areas of research include the following:

1. the use of foaming agents and other additives to reduce fingering;
2. identification of a combination of injected gases that will displace the carbon dioxide slug miscibly;
3. knowledge of the effect of injected water on the ability of carbon dioxide to contact oil in the reservoir;
4. information on reservoir conditions that affect carbon dioxide flooding both adversely and favorably (for example, the force of gravity in dipping reservoirs may operate to reduce fingering);
5. improvement of reservoir simulation models to account more accurately for the way process variables affect the performance of carbon dioxide flooding under actual reservoir conditions; and
6. improvement in the efficiency of facilities for separating and recycling carbon dioxide.

Concurrently with this research, exploration and planning activities are being conducted to obtain adequate sources of carbon dioxide at a reasonable cost.

Self-Test

5: Carbon Dioxide Flooding

(Multiply each correct answer by four to arrive at your percentage of competency.)

Fill in the Blank and
Multiple Choice with One or More Answers

1. _____ Compared to high-pressure methane, carbon dioxide –
   A. achieves miscibility on first contact with oil.
   B. can miscibly displace oil at lower pressures.
   C. usually has a lower sweep efficiency.
   D. can vaporize oil lacking methane and LPGs.
   E. has or does all of the above.

2. Describe the steps of the process by which carbon dioxide displaces oil miscibly. _____

   _____

   _____

   _____

   _____

   _____

   _____

   _____

   _____

   _____

   _____

   _____

3. Define *miscibility pressure*. _____

   _____

   _____

   _____

   _____

   _____

   _____

4. Two factors that influence the miscibility pressure required for a particular reservoir are –
   (1) _____
   (2) _____

5. About 5 percent of the crude oil is left behind during carbon dioxide flooding because _____

_____

_____.

6. State one way in which carbon dioxide displaces oil immiscibly.

_____

_____

7. The injection scheme used most widely in carbon dioxide flooding is _____

_____.

8. State one advantage and one disadvantage of injecting water during a carbon dioxide flood.

    (1) Advantage: _____

    _____

    (2) Disadvantage: _____

    _____

9. _____ When larger slugs of carbon dioxide are injected –
    A. sweep efficiency problems are overcome.
    B. the cost per barrel of oil increases.
    C. more oil is recovered.
    D. higher miscibility pressures are required.
    E. all of the above occur.

10. In laboratory tests of carbon dioxide flooding, displacement efficiency approaches _____ percent.

11. In successful field tests of carbon dioxide flooding, sweep efficiency has been about _____ percent.

12. _____ Poor sweep efficiency during carbon dioxide flooding can be caused by –
    A. reservoir heterogeneities.
    B. gravity segregation.
    C. heavy hydrocarbon residues.
    D. fingering of the carbon dioxide.
    E. all of the above.

13. _____ Reservoir conditions hindering successful use of carbon dioxide flooding include –
   A. large gas caps.
   B. oil saturation above 25 percent.
   C. chemical reaction with carbonate rock.
   D. large formation fractures.
   E. all of the above.

14. The actual fracturing pressure for a particular reservoir is determined by _____

   _____

   _____.

15. Predicting the performance of a carbon dioxide flood is difficult because the data from

   physical and computer models _____

   _____

   _____

   _____.

16. _____ The main limit on the use of carbon dioxide flooding as an improved recovery
   method is the –
   A. availability of plentiful carbon dioxide.
   B. poor sweep efficiency.
   C. depth of the well.
   D. high initial financial investment.

17. Compared to other oil-producing areas in the United States, the West Texas area is particularly suited for the use of carbon dioxide flooding because –

   (1) _____

   _____

   _____

   (2) _____

   _____

   _____

18. Carbon dioxide flooding is expected to produce about _____ percent of all the oil

   recovered by improved recovery methods.

6

Thermal Recovery

OBJECTIVES

Upon completion of this section, the student will be able to:

1. Define *thermal recovery* and describe its main use in improved recovery.

2. Explain the displacement process and mechanisms of steam drive.

3. Identify the reservoir conditions necessary for successful steam drive.

4. Define *oil-steam ratio* and explain the importance of this ratio in determining the profitability of a steam-drive project.

5. Define *steam soak* and explain the process and the displacement mechanism.

6. Define *forward in situ combustion,* identify the zones created in the reservoir during this type of combustion, and explain the displacement mechanisms occurring in each zone.

7. Identify factors that affect the amount of coke deposited, the rate of combustion, and the amount of heat loss during in situ combustion.

8. Define *wet in situ combustion* and list the advantages of this method.

9. Compare steam drive with in situ combustion in terms of recovery efficiency, production problems, and future use.

INTRODUCTION

Thermal methods of improved recovery introduce heat into a reservoir to lower the viscosity of the oil and facilitate its flow. These methods are used primarily with high-viscosity and high-density crude oils that respond poorly to other recovery methods (fig. 48).

One approach is to generate heat on the surface and transmit it to the formation in the form of steam or hot water. The injection of boiling hot water is known as *hot waterflooding;* it is less efficient than steam in heating reservoir oil and produces more water from the well. There are two types of steam injection, steam drive and steam soak. In *steam drive* or *continuous steam injection,* steam is pumped into an injection well and drives oil through the formation toward production wells. In *steam soak* or *cyclic steam injection,* steam is pumped into a production well to heat the oil in the rock surrounding the well. The well may be shut in for a period to allow heat to "soak" or distribute evenly before the well is produced.

Another approach, known as *in situ combustion* or *fire flooding,* is to generate heat within a reservoir by injecting air and burning a portion of the oil in place. The most common variant, *forward combustion,* involves starting a fire in the formation near an injection well and injecting air into the well. The fire and the air move together toward production wells. In a rarer variant known as *reverse combustion,* the path of the fire moves from production to injection well, counter to the flow of injected air rather than in the same direction. There are two types of forward combustion, dry and wet. When air is the only injected heat carrier, the method is called *dry combustion.* Often, however, water is injected simultaneously or alternately with air in order to reduce heat losses and regulate the burning process. When water is used, the method is known as *wet combustion* or *COFCAW* (combination of forward combustion and waterflooding).

STEAM DRIVE

Steam drive is currently the most widely used improved recovery method, excluding waterflooding. Like waterflooding, it uses a pattern of injection and production wells, except that the wells are likely to be more closely spaced. Steam of 80 percent quality is commonly used, meaning that 80 percent of the water by weight is converted to steam, and the other 20 percent is hot water at the boiling point.

In general, steam drive is superior to hot waterflooding because the heat content of a pound of steam is higher than that of a pound of water at the same temperature and pressure. Moreover, steam can usually be injected at a faster rate and less water is produced when steam is used. However, hot waterflooding has been used successfully under conditions unfavorable for steam drive.

Displacement Mechanisms

In the steam drive method, the main mechanism for oil recovery is the reduction of oil viscosity by heat. Other contributing mechanisms include (1) expansion of the oil, (2) miscible displacement of the oil by distilled light hydrocarbons, (3) changes in the relative permeabilities to oil and water, (4) gas (steam) drive, and (5) hot water drive. The importance of these mechanisms varies with differences in reservoir conditions and characteristics of the crude oil. The contribution of each mechanism to the

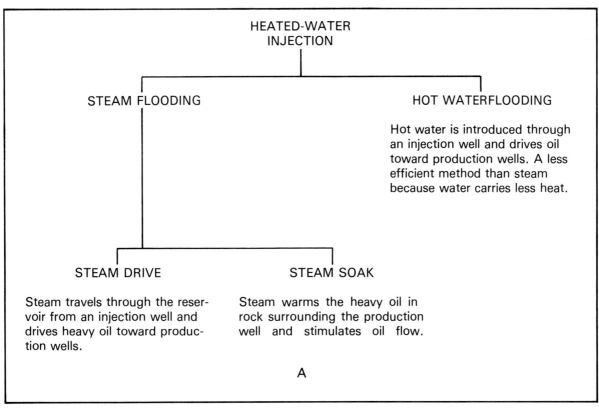

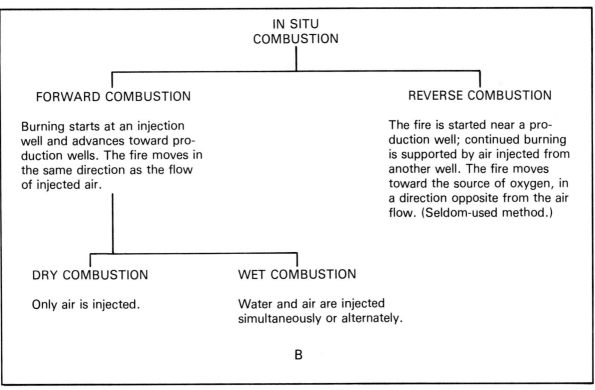

Figure 48. Thermal recovery processes. *A*, heat generated on the surface; *B*, heat generated in the reservoir. Researchers are seeking ways to reduce heat loss by generating steam downhole.

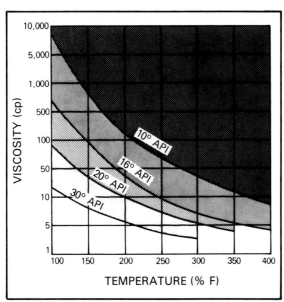

Figure 49. Viscosity of four oils at various temperatures

performance of the project may be difficult to determine exactly.

The greatest reduction of oil viscosity occurs at the first increases in temperature (fig. 49). Since the chart in figure 49 is based on a logarithmic scale, the distance of a number from the starting point represents its logarithm rather than its size. As a result, a short distance represents a large number. For example, on the curve representing a 10° API oil, the chart shows that heating from 100°F to 150°F reduces the viscosity of the oil from about 9,000 centipoise (cp) to about 800 cp. Heating the 10° API crude oil to 300°F reduces viscosity to about 35 cp (table 9). The greatest reduction of viscosity takes place during the first 200° of heating—that is, during the temperature increase from 100°F to 300°F. Heating affects higher-gravity oils less dramatically, as shown in figure 49 by the flatter viscosity curves for the 20° and 30° API crude oils.

Heat has several additional effects: expansion of the oil, vaporization of lighter hydrocarbons, and expansion of the formation rock. Oil expansion provides additional energy to move oil from the reservoir. The vaporized hydrocarbons move ahead of the steam zone, cooling and condensing into liquid solvents that can dissolve and displace crude oil. There is some evidence that heating the rock increases its relative permeability to oil. *Relative permeability* is the ease with which a fluid flows through a rock formation when the rock is saturated with two or more fluids such as oil, gas, and water. An increase in relative permeability to oil would permit the oil to flow more easily than the water.

During a steam drive, oil is moved along by gas and water drive. The steam provides gas drive. As it moves further into the formation and cools, it condenses to hot water and provides water drive.

TABLE 9
Effect of Increasing Temperature on Viscosity
(10°API Crude Oil)

Temperature (°F)	Viscosity (cp)
100	9,000
150	800
200	200
250	65
300	35

Process Description

Drive zones. Under ideal reservoir conditions, steam drive forms three concentric zones as it progresses from injection to production wells: a steam zone, a hot water zone, and an oil bank (fig. 50). Ahead of these is a zone of water and oil not yet affected by the steam drive. The steam zone forms around the injection well and expands as injection continues. Its

90

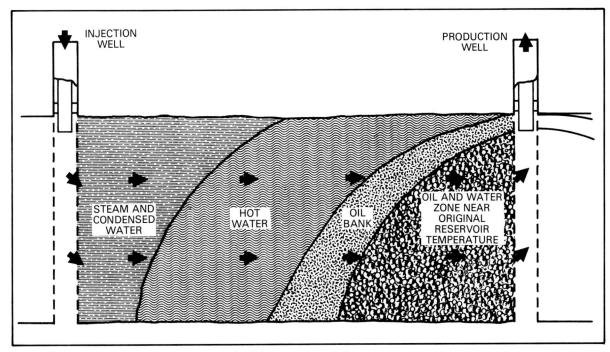

Figure 50. Steam drive displacement of oil from a reservoir

temperature is about equal to the heat of the injected steam. In the steam zone, oil is displaced by vaporization of lighter hydrocarbons and their condensation into liquids that dissolve crude oil. The steam also provides a certain amount of gas drive. As the steam zone advances and contacts cooler portions of the reservoir, it condenses to hot water. Temperatures in the hot water zone decrease from the temperature of the steam at the back boundary to the temperature of the reservoir at the front boundary. In the hot water zone, the oil is displaced through swelling, viscosity reduction, and hot waterflooding. A portion of the oil is accumulated in a bank ahead of the hot water. These zones progress until they reach the production wells.

Under actual reservoir conditions, the pattern of oil displacement is likely to be more complex. Because of its high mobility, steam has a tendency to form a direct channel to production wells, banking oil and water in its path. Steam injected after breakthrough will spread at a lower pressure and a slower rate into the upper part of the reservoir. Steam condensing into water will travel downward into the lower portion of the reservoir. In some reservoirs, a substantial portion of oil is produced after breakthrough to the production wells.

Design problems. Loss of heat and gravity segregation are the two most serious problems in designing a steam drive project. Because the reduction of oil viscosity by heat is the main displacement mechanism, the success of a project depends on the amount of heat transferred to the crude oil rather than to the formation. Gravity segregation of steam and hot water may substantially reduce sweep efficiency.

The temperature required to boil water rises from 212°F at the atmospheric pressure of 14.7 pounds per square inch (psi) to around 600°F at 1,500 psi. The latter pressure is the maximum at which most steam drive projects operate and corresponds to a depth of about 2,500 feet. Due to heat losses in transmission, the temperature of steam reaching the formation in most projects ranges from about 350°F to somewhat over 450°F. These heat losses occur from surface lines and from the wellbore. Surface heat loss can be reduced to about 10 percent by insulating or burying steam lines. Losses in the wellbore can be reduced to some extent by special well completions that insulate the tubing.

The greatest heat loss occurs in the reservoir. Heat is conducted through the rock above and below the steam zone and lost from the oil-bearing rock. When the pay zone is thicker, the percentage of heat loss is less (fig. 51). To contact a given number of acre-feet in a thin formation (for example, a thickness of 10 feet), steam must spread out over a wider area than in a thick formation. The steam is exposed to a larger surface area of rock surrounding the formation and loses more heat. In a thicker pay zone (for example, 50 feet) steam does not have to spread as far to heat the same reservoir volume, and more than half the heat remains within the oil-bearing layer. Therefore, engineers follow guidelines for minimum pay-zone

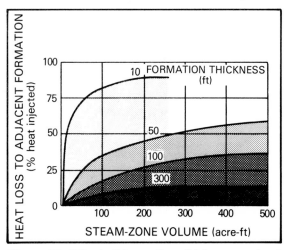

Figure 51. Effect of formation thickness on heat loss

thickness in deciding whether to apply steam drive to a reservoir.

In the operation of a steam drive project, heat can be conserved by spacing wells more closely, increasing injection rates, and switching from steam drive to waterflood. Spacing wells more closely shortens the distance the steam must travel for breakthrough and allows less time for heat loss to adjacent rock. Wells may be spaced as closely as 1 well per acre. More rapid injection rates speed the progress of the steam flood through the formation and also allow less time for heat loss through conduction. After the steam zone is about halfway from injection to production wells, steam injection can be discontinued and cold water used to complete the project. The water is heated by the hot rock and the remaining steam, and it carries this heat forward as a hot waterflood. Fuel used for firing steam generators is also reduced. The change to waterflood should be planned to occur at a point when completing floodout with steam would be uneconomic—that is, when the cost of fuel needed to continue firing the steam generators would be more than the value of the additional oil recovered.

Another design problem is gravity segregation of steam and hot water in the reservoir. In most projects, the injected fluid is about 80 percent steam and 20 percent hot water. The steam, being lighter in density than water, tends to override the water and to travel into the upper part of the reservoir, and the water tends to drop into the lower part. Since hot waterflooding is not as efficient as steam in displacing oil, a steam drive project may leave behind twice as much oil in the lower parts of the reservoir as in the higher reaches. This problem can be reduced to some extent by injecting the steam as low as possible in the pay zone.

Range of Applications

Steam drive is suited to the recovery of viscous, low-gravity crude oils. Heat is most effective in reducing the viscosity of crude oils with gravities in the range of 10° to 20° API. Oil viscosity in successful steam drive projects has ranged initially from 400 to 4,000 cp. High viscosity and low gravity (high density) do not necessarily coincide. Two fluids of the same gravity may have different viscosities because of differences in their chemical compositions. Generally, however, heavy oils—those with a low-gravity API number—are also highly viscous.

Minimal saturation should range between 50 and 60 percent, according to U.S. Department of Energy studies of fields using steam drive. Many steam drive projects use some of the recovered oil as fuel in generating injection steam. The higher the oil saturation, the higher is the oil recovery per barrel of injected steam. Oil recovery must exceed the use of oil as steam generator fuel by a margin great enough to make the project economical.

Steam drive can be applied to pay zones with depths up to about 5,000 feet. Pressures beyond that depth require too much fuel to generate steam at the necessary temperatures. Also, heat losses in the wellbore escalate at greater depths. The minimum depth for operation of steam drive is about 200 feet so that injection can operate at a high enough pressure to ensure an economic rate of steam-zone advance through the reservoir. If the steam advances too slowly, heat loss to adjacent formations rises unacceptably. In California, steam drive projects have generally operated between depths of 500 and 2,000 feet.

A sufficient rate of steam-zone advance also depends upon permeability.

Authorities suggest that a formation have a minimum permeability of 500 millidarcys (md). Certain California steam drive projects have operated in reservoirs with permeabilities of over 2,000 md.

To minimize heat loss, the use of steam drive should be limited to producing formations over 20 feet thick and preferably over 30 to 50 feet thick.

Advantages and Limitations

Steam drive is a proven method for improved recovery from heavy-oil reservoirs. When steam drive can be used, U.S. operators prefer it to in situ combustion for several reasons. Steam drive has a faster response rate and a higher recovery efficiency: 25 to 64 percent of residual oil in place as compared to 28 to 39 percent for in situ combustion. Steam drive operation is somewhat simpler and less likely to damage wells and equipment. Since it does not involve combustion and hydrocarbon cracking, it does not create noxious combustion gases and highly corrosive water.

Steam drive cannot be used in deep reservoirs, thin formations, or formations with low permeability. The high-quality water required to produce 80 percent quality steam may be difficult to acquire in some areas, and water treatment adds to production costs. Successful operation requires continual monitoring activity and special solutions to equipment and production problems. Cement failure often occurs in conventionally completed wells. New wells must be equipped for high-temperature operation. Emulsions formed by heavy crude oils and water clog wellbores and equipment and require costly treatments to disperse.

Operation at high temperatures increases safety risks and requires special training of personnel. Production of hot fluids can heat wellheads and gathering lines. Safety measures include burying lines and fencing wellheads. State and local regulations may require 24-hour monitoring of large steam-generation facilities. The heat and noise of steam recovery can become an environmental concern in areas where urban development encroaches on established oil fields.

Economic and Technical Outlook

Costs. The ratio of barrels of recovered oil to barrels of injected steam establishes the economic limits of steam drive. Authorities consider an *oil-steam ratio* of around 0.25 to 1 to be favorable for economic success.

Steam is produced by burning natural gas or crude oil in steam generators. Often, part of the produced crude oil is used as steam-generator fuel. One barrel of oil can generate about 13 to 14 barrels (bbl) of steam. In a steam-drive project with an oil-steam ratio of 0.25 to 1, 1 bbl of every 3⅜ bbl of produced oil is used as fuel for generating steam. Four barrels of steam are required, in turn, to produce 1 bbl of oil (fig. 52). Generally, steam drive projects consume as fuel 1 of every 3 to 4 bbl of oil produced.

The value of the recovered oil must cover the cost of fuel for the steam generators as well as other project costs such as operation, maintenance, taxes, and profits. Therefore, every project has an oil-steam ratio below which further steam injection ceases to be economical. For example, in a steam drive project in California, an oil-steam ratio of 0.12 was determined to be the economic cutoff point. At that ratio, more than half the oil recovered would be consumed in steam generation. The oil remaining for marketing would not be enough to pay for

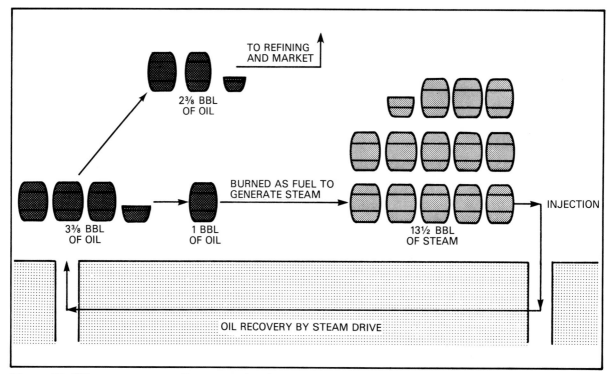

Figure 52. Oil recovery at an oil-steam ratio of 0.25 to 1. Four bbl of steam are needed to recover 1 bbl of oil. Of every 3 3/8 bbl of oil recovered, 1 bbl must be burned to generate steam for continued oil recovery.

94

other project costs. In projects burning lease crude oil, the mid-1980 cost of producing a barrel of oil ranged from $21 to $28. If, instead, a project purchased crude oil for its steam generators, the price of a barrel of oil would rise to $27 to $35, and the generation of steam would account for about half the total cost.

Recovery potential. At the end of the 1970s, steam drive was more highly developed than other improved recovery methods and had economically recovered considerable amounts of oil. In 1977, steam drive recovered 140,000 barrels per day (bbl/d), or 53 percent of the improved recovery total of 265,000 bbl/d. (Not counted in this total was 120,000 bbl/d recovered by steam soak.)

In the future, steam drive will account for a slightly smaller portion of improved recovery as other methods are improved. Nevertheless, steam drive will account for one-quarter to one-third of all future improved recovery; estimates for steam drive recovery range from 4 to 17 billion barrels. The National Petroleum Council study predicts that California heavy-oil fields will produce 80 percent of U.S. oil recovered by steam drive.

Recent pilot projects in South Texas also look promising; they have been using fracture-assisted steam technology (FAST), the use of fracturing to increase formation permeability before injecting steam. The fracture patterns are small to prevent channeling. Fracturing permits much faster production of heavy crude oils.

Technical advances. Although steam drive is a well developed method, it can be improved by finding ways to increase recovery efficiency and to reduce production costs. Researchers are experimenting with foams, gels, and polymers to reduce steam channeling and with chemicals to increase displacement in the hot water zone. Better information about the effects of steam quality and injection rate on performance may lead to more effective project designs.

Reduction of heat loss is an important improvement. One technique under consideration is downhole steam generation, which would eliminate surface and wellbore heat loss and might extend the use of steam drive to reservoirs deeper than the current 3,000-foot depth limit. Another possible advantage would be the discharge of combustion gases into underground rock rather than into the air. This capability would eliminate the air pollution problem created by in situ combustion. However, in field tests to date, downhole steam generators have been unsuccessful because of severe corrosion problems. The cost of generating steam can be reduced by using cheaper fuels such as coal. Ways are also being sought to extend the use of steam drive to reservoirs with tar sands or with lighter oils.

STEAM SOAK

Steam soak, also known as *cyclic steam injection* and *huff 'n' puff,* consists of the injection of steam into a production well in order to increase the flow of heavier oils into the well. It is considered by some to be a well stimulation method because it is used to increase production from individual wells rather than to introduce artificial drive into the reservoir through injection wells.

Steam soak begins with the injection of 5 to 15 thousand barrels of steam into a production well to heat the immediate area. After this initial injection, production may begin right away and proceed for several

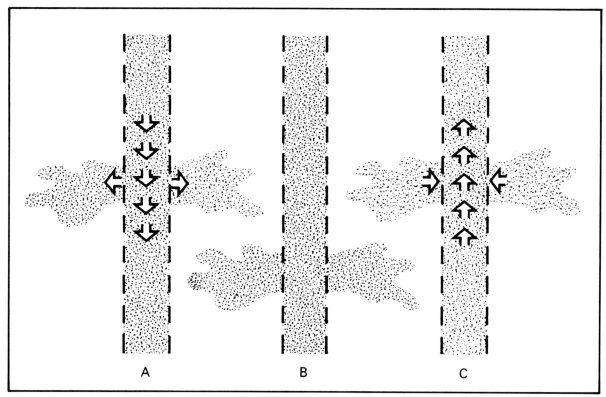

Figure 53. Steam soak method of stimulating the flow of heavy oil from production wells. *A,* steam injection; *B,* steam soak; *C,* oil and water production.

months to a year (fig. 53). Or the well may be shut in for a period of time, allowing it to "soak" before it goes back on production. The hot fluid causes displacement primarily by lowering the viscosity of the oil but may also boost production by removing deposits in the well. Production increases in the initial cycle may reach several times the well's previous yield.

There may be as many as 20 cycles of injection, soak, and production. Response is best in steeply dipping reservoirs; as the heat spreads and causes viscous oils to flow more easily, they drain into the production wells under the influence of gravity. Eventually, response declines, and further injections of steam will not bring profitable yields of oil. At this point, the wells may benefit from the application of another method such as steam drive.

IN SITU COMBUSTION

In situ combustion is a method of improved recovery in which heat is generated within the reservoir by injecting air and burning a portion of the oil in place. Injected air keeps the subsurface fire burning the way a cigar ash burns. The amount of oil burned can be controlled by the amount and rate of air injection. As the heavier hydrocarbons in the crude oil are consumed by combustion, the fire front moves on to a new area. The lighter hydrocarbons are mobilized and produced.

Displacement Mechanisms

During in situ combustion, several mechanisms are responsible for the displacement of oil. The most important is the reduction of oil viscosity by heat. Other

96

mechanisms include (1) thermal cracking of hydrocarbons and vaporization of the lighter hydrocarbons; (2) gas drive from steam, combustion gases, and vaporized hydrocarbons; (3) miscible displacement by condensed lighter hydrocarbons; and (4) hot water drive. Because the mechanisms differ in each of the zones created by the fire flood, they can be best understood by examining the process at work within each of these zones.

Process Description

When oxygen in injected air contacts reservoir oil, a chemical reaction called oxidation takes place. Often, the initial temperature of the reservoir and the oxidation reaction provide enough heat to ignite reservoir oil spontaneously. If spontaneous ignition does not occur, the oil can be ignited by heating the injected air, by using a downhole heater, by injecting an oxidizable chemical such as linseed oil, or by injecting steam with the air. Steam not only increases the reservoir temperature to ignition point, but it also cleans well perforations, and it prevents well burnout from high combustion temperatures by lowering oil saturation in the rock surrounding the well.

Forward combustion. Forward combustion is the most common procedure used. The reservoir oil is ignited near an injection well, and air is injected into the well. The burning front moves toward production wells, in the same direction as the flow of injected air. Several zones are created during the combustion process: a burned zone, a flame front, a vaporizing zone, a condensing zone, and an oil bank. Ahead of these is the unaffected portion of the reservoir (fig. 54).

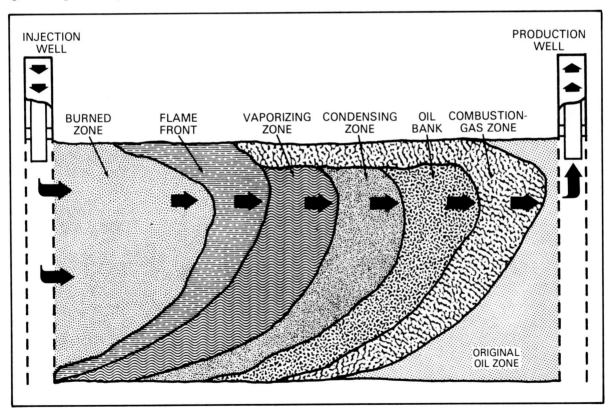

Figure 54. In situ forward combustion

97

As the burning front moves away from the injection well, a burned zone forms, consisting of clean fine-grained sand. A great portion of the heat generated remains there.

At the flame front, temperatures are highest, ranging from 600° to 1,200°F. Much action occurs in this zone. Combustion heat cracks crude hydrocarbons into smaller, lighter molecules. It vaporizes the lighter hydrocarbons. The heavier hydrocarbons are deposited as a dry residue called coke, which catches fire and becomes the fuel for the flame front. The amount of coke deposited depends on the hydrocarbon composition of the crude oil. Oxygen is used up in the combustion reaction, and hot combustion gases are released. These gases include water in the form of steam, carbon monoxide, carbon dioxide, hydrogen sulfide, and sulfur dioxide. Nitrogen, which entered with the injected air, is also present. The heat of combustion vaporizes connate water into steam. As the coke is consumed,

the flame front moves forward, progressing slowly through the reservoir at a rate of one to several inches a day and migrating toward the upper part of the reservoir. The rate and efficiency of combustion depend on the rate of air injection.

Steam, hot combustion gases, and vaporized hydrocarbons move ahead of the flame front, forming a vaporizing zone. Temperatures in this zone range from the temperature of the flame front to the temperature of boiling water, and vary according to depth and pressure (fig. 55). Steam lowers the viscosity of the oil, causing most of the displacement in this zone. Hot gases and steam also contribute some measure of displacement energy by gas drive.

As steam and hydrocarbon vapors move forward into the reservoir, they cool, forming a condensing zone. Light hydrocarbons condense into liquid solvents that miscibly displace reservoir oil. Condensed steam becomes a hot waterflood, which settles

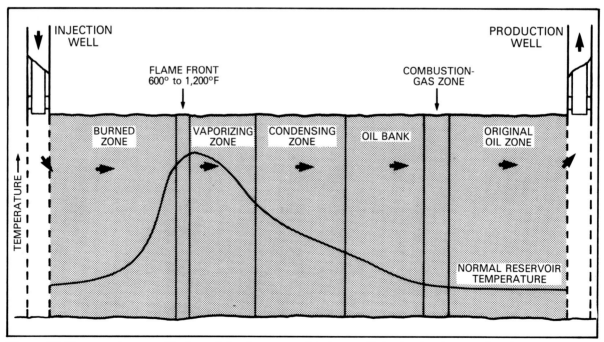

Figure 55. Temperature profile from injection to production well during in situ combustion

98

into the lower portion of the zone. In the upper portion, gases concentrate and provide gas-drive energy for displacement. Oil viscosity is lowered by heat from the liquids and gases.

Oil accumulates in a bank ahead of the condensing zone, along with combustion gases and connate water. Beyond this zone lies the portion of the reservoir not yet affected by the combustion process. Temperatures in these zones are close to original reservoir temperatures and do not benefit from viscosity reduction by heating.

Reverse combustion. In another variant of in situ combustion—*reverse combustion*—the combustion front moves in a direction opposite to that of the injected air. Air is injected into a production well, and a fire is started near the well. When the combustion zone has advanced a short distance outward from the well, air injection is switched from the production well to an adjacent injection well. The fire advances toward its source of oxygen, the injection well; but the oil moves toward the production well. This method can be used to produce very viscous oils. It can also be used to preheat the area surrounding production wells so that viscous oil will be produced more efficiently during forward combustion.

Design problems. As with steam drive, heat loss and gravity segregation are major design problems. In *dry combustion,* air is the primary transmitter of heat in the reservoir. Because air is an inefficient heat carrier, most of the heat generated by combustion stays in the burned zone and is lost by conduction to rock layers above and below the oil-bearing layer. The advancing flame front carries forward only about 20 percent of the heat to areas where it is put to work displacing oil. Because of gravity differences, vapors, steam, and gases tend to rise into the upper portion of the reservoir, and condensed liquids tend to drop into the lower portion. The result of gravity segregation is a substantial loss of sweep efficiency in the lower region of the reservoir.

To recover and transmit a larger portion of the generated heat, engineers have developed a variation of forward combustion called *wet combustion,* or *COFCAW* (combination of forward combustion and waterflooding). In this method, water is injected either simultaneously or alternately with air. When the water flows through the burned region, it absorbs heat, vaporizes to steam, and carries heat forward through the flame front. Arriving at the vaporizing zone, the steam increases the amount of viscosity reduction. Traveling on, it reaches the production wells, warming the surrounding rock and facilitating the flow of oil. Water also reduces to as little as one-third the amount of injected air required and speeds the progress of the flame front. (The reasons for this reduction are not known precisely.) Savings in the cost of compressing and injecting air may be substantial.

Gravity segregation can be reduced by closer well spacing and starting the fire as low as possible in the pay zone. Wet combustion can also help minimize this problem if air is injected into the bottom part of the pay zone and water into the top part.

Range of Applications

In situ combustion is used mainly to recover viscous, low-gravity oils, but it can also be used to recover lighter oils up to 40° API.

In determining whether in situ combustion is suitable for a particular reservoir, engineers consider guidelines for fuel content, air requirement, depth and thickness of the reservoir, oil saturation, and permeability.

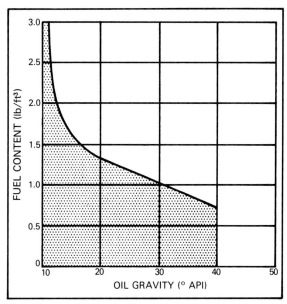

Figure 56. Effect of oil gravity on fuel content

The *fuel content* of a reservoir crude oil—the amount of coke it deposits for the flame front—is the most important consideration in deciding whether in situ combustion can be applied. The oil must deposit enough fuel to keep the flame front burning, but too much fuel slows the movement of the front through the reservoir, because the fire consumes all available fuel before advancing. As a result, the cost of air injection rises, and heat loss increases. Fuel content is measured in pounds per cubic foot (lb/ft³) of reservoir rock. Reported fire floods have operated at less than 1 to just under 3 lb/ft³. Fuel content correlates with the API gravity of crude oil (fig. 56); lower-gravity oils have a higher fuel content. Engineers designing a fire flood can get a laboratory analysis showing the actual fuel content of a particular crude oil.

The air requirement is determined by the amount of fuel deposited and correlates with API gravity (fig. 57). Heavier crude oils require more air. The air requirement is measured in million standard cubic feet per

acre-foot (MMscf/acre-ft) under standard conditions of 60°F and atmospheric pressure.

Deep reservoirs do not present technical limitations—one in situ combustion project operated at over 11,000 feet. However, air compression costs rise with depth, and most reported projects have worked at fewer than 5,000 feet. To retain enough heat for efficient performance, the pay zone must be at least 10 feet thick.

Because a portion of reservoir oil is consumed during in situ combustion, oil saturation must be high enough to permit recovery of a profitable amount of oil. Many authorities consider a residual saturation of 50 percent to be the economic minimum. In a Louisiana project studied for the U.S. Department of Energy, researchers estimated that an initial oil saturation of 69

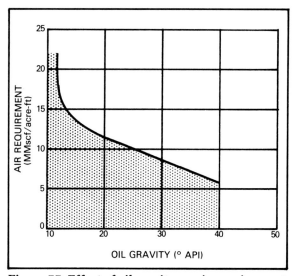

Figure 57. Effect of oil gravity on air requirement

percent would result in a recovery of 28 percent of the oil in place (table 10). Oil saturation may also be expressed in barrels per acre-foot (bbl/acre-ft). In the early 1980s, a saturation of 780 bbl/acre-ft was considered a minimum for economic success.

TABLE 10
Oil Consumption and Recovery
in the Caddo Pine Field

Reservoir Factors	Barrels	Percentage
Oil saturation at start of project	295,000	69 (original oil in place)
Recovery	84,000	28 (residual oil in place
Amount of fuel burned	36,000	12 (residual oil in place)

SOURCE: Lewin and Associates, Inc., *Economics of Enhanced Oil Recovery, Final Report.* DOE/ET/12072-2. Washington, D.C.: U.S. Department of Energy, 1981.

Reservoir rock must have a permeability of more than 100 md to allow heavy, viscous oil to flow. In combination with other reservoir data such as formation thickness and oil viscosity, the permeability also indicates whether the burning front will advance rapidly enough to prevent excessive heat loss.

Advantages and Limitations

Although mainly used to recover viscous heavy oil, in situ combustion can also be used with lighter oils ranging in gravity up to 40° API. Air is an easily obtainable injection fluid, compared to the high-quality water used in steam injection. Depth of operation is not limited technically, as with steam injection. Poor heat distribution can be overcome to some extent by using wet combustion.

In situ combustion projects encounter numerous production problems because of the heat and the combustion products generated by fire. Heat-resistant cement and downhole equipment must be used to complete wells. Corrosion from sulfurous combustion gases and from carbon dioxide is reduced by using equipment made of corrosion-resistant materials and by applying corrosion treatments. Injection and production rates are lowered when wells become clogged with scale, emulsions, asphaltene, carbon, and sand. Sand erosion increases wear on tubing and pumps. When the flame front arrives at production wells, they must be protected from high temperatures by circulating water through the annular space.

In situ combustion also poses safety and environmental problems. The injection of air creates a potential explosion hazard. Operation of air compressors generates high noise levels. Sound-reducing materials must be used to confine noise to the immediate area, and operating personnel require ear protection. Combustion produces poisonous gases—carbon monoxide and hydrogen sulfide—that are hazardous to operating personnel and to the environment. Air pollution is a serious disadvantage of this method.

Economic and Technical Outlook

Costs. According to the U.S. Department of Energy, the cost of a barrel of oil recovered by in situ combustion ranges from $25 under favorable conditions (if the flame front and operating problems can be controlled) to $36 in more difficult circumstances (if, for example, the pay zone is at a greater depth). However, costs may be considerably lower in states with less strict environmental regulations, in large fields, and in projects that do not require much testing.

In a typical in situ combustion project, the compressor is the most expensive investment item. The highest annual costs are for fuel and compressor operation and maintenance. Consequently, the ratio of injected air to recovered oil mainly determines whether an in situ combustion project can operate economically. Successful projects have run at air-oil ratios of less than 20 Mscf/bbl.

101

Recovery potential. In situ combustion trails steam drive in both efficiency and expected total recovery. The U.S. Department of Energy study has predicted that in situ combustion will contribute from 1.6 to 2.0 billion barrels of oil to U.S. recovery, compared to a recovery of 4 to 17 billion barrels by steam drive. According to this forecast, in situ combustion will account for 11 to 29 percent of thermal recovery, and for 4 to 9 percent of recovery by all improved recovery methods. These relatively low figures reflect the preference of the U.S. oil industry for steam drive rather than in situ combustion when a thermal method is applicable. A mid-1970s study of the National Petroleum Council found that in California heavy-oil fields, steam drive was preferred for reservoirs above 5,000 feet. California has two-thirds of the nation's heavy oil, and two-thirds of this oil is located above 3,000 feet. The preference for steam drive was also confirmed by a 1980 survey of improved recovery project operators. The survey included 100 thermal recovery projects; of these, only 9 used in situ combustion. Although 7 of the 9 were rated successful, they were outnumbered almost ten to one by the 65 steam drive projects that received the same rating.

Recovery from tar sands is probably the greatest future application of in situ combustion. Tar sands contain heavy hydrocarbons so viscous that they will not flow at reservoir temperatures. They have the immobility of a solid. Mining has been proposed to recover the uppermost deposits — those covered by no more than 250 feet of overburden. The deeper deposits are a target for recovery by in situ combustion.

One of the world's largest oil reserves, estimated at over 800 billion barrels, lies in the Alberta, Canada tar sands. Only about 10 percent can be mined. Since the 1950s, a Canadian oil company has been testing all types of thermal recovery to determine the most effective way to recover oil from tar sands. A three-step design has been found most effective. After a period of dry forward combustion, air injection is stopped and time is allowed for the heat to be distributed evenly. Then wet combustion is begun to displace the heated tar. The produced tar must be upgraded before transportation. One proposed method is extraction of the heaviest hydrocarbons by distillation. They can be used to provide fuel for the upgrading facility. Large-scale production will be needed to justify the cost of such a facility — about 100,000 bbl/d. At the earliest, commercial production will begin in the 1990s. Other influences, such as world oil prices and Canadian national tax policies, add financial uncertainties to future recovery from tar sands, but the potential remains enormous.

Technical advances. The recovery potential of in situ combustion can be increased by technical advances in the following areas:

1. ways to increase vertical sweep efficiency;
2. identification of conditions favorable to in situ combustion, such as shallow formations;
3. improvement of the wet combustion process to reduce air-oil ratios; and
4. development of more efficient fire flood techniques and designs for recovery of oil from tar sands.

Self-Test

6: Thermal Recovery

(Multiply each correct answer by three and add one bonus point to arrive

at your percentage of competency.)

Fill in the Blank and
Multiple Choice with One or More Answers

1. Thermal recovery methods are generally used to recover oils with the following characteristics:

   (1) _____

   (2) _____

2. Steam drive is usually more efficient than hot waterflooding because _____

   _____

   _____

   _____

   _____.

3. _____ In the steam zone formed around an injection well, oil is displaced by –
   A. thermal cracking.
   B. vaporization of hydrocarbons.
   C. gas drive.
   D. miscible displacement.
   E. all of the above.

4. Two zones in advance of the steam zone are –

   (1) _____

   (2) _____

5. _____ The amount of heat lost from a reservoir is dependent on –
   A. formation thickness.
   B. rate of steam drive advance.
   C. oil saturation.
   D. formation depth.
   E. all of the above.

6. The purpose of injecting steam as low as possible in the pay zone is _____

   _____

   _____.

7. The purpose of converting a steam drive project to a waterflood is _____

_____

_____.

8. Define *oil-steam ratio*. _____

_____

9. Explain why the oil-steam ratio is important in determining the profitability of a steam drive project. _____

_____

_____

_____

_____

10. State one safety or environmental problem caused by steam drive. _____

_____

11. Steam soak causes oil to flow into the production well primarily by _____

_____.

12. A fire flood is spontaneously ignited by the heat from –

(1) _____

(2) _____

13. Explain how coke is formed. _____

_____

_____

_____

14. The higher the API gravity of a crude oil, the _____(greater/less) is the amount of coke deposited.

15. Define *forward combustion*. _____

_____

_____

_____

_____

104

16. _____ During forward combustion, oil is displaced in the condensing zone by –

   A. gas drive.
   B. viscosity reduction.
   C. hydrocarbon solvents.
   D. hot waterflood.
   E. all of the above.

17. _____ In dry combustion, most of the heat is located in the –

   A. condensing zone.
   B. oil bank.
   C. vaporizing zone.
   D. burned zone.

18. Factors that affect the rate of combustion include –

   (1) _____

   (2) _____

19. Wet combustion is often preferred to dry combustion because –

   (1) _____

   _____

   (2) _____

   _____

20. _____ Closer well spacing is used to –

   A. reduce heat loss.
   B. improve sweep efficiency.
   C. improve displacement efficiency.
   D. reduce the injection-fluid requirement.
   E. accomplish all of the above.

21. Name one safety or environmental problem caused by in situ combustion. _____

   _____

   _____

22. _____ In the future, in situ combustion will probably be used with –

   A. crude oils with gravities up to 60° API.
   B. extremely viscous crude oils in tar sands.
   C. formations with less than 250 feet of overburden.
   D. reservoirs with oil saturations of 20 percent or more.
   E. all of the above.

23. The oil industry often prefers to use steam drive rather than in situ combustion because –

(1) _____

_____

(2) _____

_____

Glossary

A

absolute permeability *n:* a measure of the ability of of a single fluid (such as water, gas, or oil) to flow through a rock formation when the formation is totally filled (saturated) with the single fluid. The permeability measure of a rock filled with a single fluid is different from the permeability measure of the same rock filled with two or more fluids. Compare *relative permeability* and *effective permeability*.

absorption *n:* the taking up on one substance in bulk by another substance, as, for example, the dissolving of gas in oil. Compare *adsorption*.

acidize *v:* to treat oil-bearing limestone or other formations with acid for the purpose of increasing production. Hydrochloric or other acid is injected into the formation under pressure. The acid etches the rock, enlarging the pore spaces and passages through which the reservoir fluids flow. Acidizing is also used to remove formation damage by dissolving material plugging the rock surrounding the wellbore. The acid is held under pressure for a period of time and then pumped out, after which the well is swabbed and put back into production. Chemical inhibitors combined with the acid prevent corrosion of the pipe.

acre-ft *abbr:* acre-foot.

adhesion *n:* the tendency, due to intermolecular forces, for molecules of one substance to cling to those of another substance.

adsorption *n:* the adhesion of a thin film of gas or liquid to the surface of a solid. Liquid hydrocarbons are recovered from natural gas by passing the gas through activated charcoal, silica gel, or other solids, which extract the heavier hydrocarbons. Steam treatment of the solid removes the adsorbed hydrocarbons, which are then collected and recondensed. The adsorption process is also used to remove water vapor from air or natural gas. Compare *absorption*.

alkaline (caustic) flooding *n:* a method of improved recovery in which alkaline chemicals such as sodium hydroxide are injected during a waterflood or combined with polymer flooding. The chemicals react with the natural acid present in certain crude oils to form surfactants within the reservoir. The surfactants enable the water to move additional quantities of oil from the depleted reservoir. See *micellar-polymer flooding*.

annular space *n:* 1. the space surrounding a cylindrical object within a cylinder. 2. the space around a pipe in a wellbore, the outer wall of which may be the wall of either the borehole or the casing; sometimes termed the annulus.

anticlinal trap *n:* a hydrocarbon trap in which petroleum accumulates in the top of an anticline. See *anticline*.

anticline *n:* an arched, inverted-trough configuration of folded and stratified rock layers. Compare *syncline*.

API gravity *n:* the measure of the density or gravity of liquid petroleum products in the United States, derived from specific gravity in accordance with the following equation:

$$\text{API gravity} = \frac{141.5}{\text{specific gravity}} - 131.5$$

API gravity is expressed in degrees, 10° API being equivalent to 1.0, the specific gravity of water.

artificial lift *n:* any method used to raise oil to the surface through a well after reservoir pressure has declined to the point at which the well no longer produces by means of natural energy. Sucker rod pumps, gas lift, hydraulic pumps, and submersible electric pumps are the most common forms of artificial lift.

B

bbl *abbr:* barrel.

bbl/acre-ft *abbr:* barrels per acre-foot.

bbl/d *abbr:* barrels per day.

bottom water *n:* water found below oil and gas in a producing formation.

bubble point *n:* 1. the temperature and pressure at which part of a liquid begins to convert to gas. For example, if a certain volume of liquid is held at constant pressure, but its temperature is increased, a point is reached when bubbles of gas begin to form in the liquid. That is the bubble point. Similarly, if a certain volume of liquid is held at a constant temperature but the pressure is reduced, the point at which gas begins to form is the bubble point. Compare *dew point*. 2. the temperature and pressure at which gas, held in solution in crude oil, breaks out of solution as free gas.

C

capillarity *n:* the rise or fall of liquids in small-diameter tubes or tubelike spaces, caused by the combined action of surface tension and wetting. See *capillary pressure.*

capillary pressure *n:* a pressure or adhesive force caused by the surface tension of water. This pressure causes the water to rise higher in small capillaries in the formation than it does in large capillaries. Capillary pressure in a rock formation is comparable to the pressure of water that rises higher in a small glass capillary tube than it does in a larger tube.

carbonate rock *n:* a rock composed principally of carbonates (a salt of carbonic acid, H_2CO_3), especially if it is at least 50 percent carbonates by weight. Limestone and dolomite are two carbonate rocks that sometimes contain petroleum.

cementation *n:* the precipitation of a binding material around grains or minerals in rocks.

centipoise *n:* one-hundredth of a poise.

chemical flooding *n:* See *micellar-polymer flooding* and *alkaline (caustic) flooding.*

chemical treatment *n:* any of many processes in the oil industry that involve the use of a chemical to effect an operation. Some chemical treatments are acidizing, crude-oil demulsification, corrosion inhibition, paraffin removal, scale removal, drilling fluid control, refinery and plant processes, cleaning and purging operations, chemical flooding, and water purification.

cm *abbr:* centimetre.

COFCAW *abbr:* combination of forward combustion and waterflooding. Also called wet combustion. See *wet combustion.*

cohesion *n:* the attractive force between the same kinds of molecules (i.e., the force that holds the molecules of a substance together).

coke *n:* 1. a solid cellular residue produced from the dry distillation of certain carbonaceous materials, containing carbon as its principal constituent. 2. a residue of heavier hydrocarbons formed by thermal cracking and distillation and deposited in the reservoir during in situ combustion. This residue catches fire and becomes the fuel for continued combustion.

combination drive *n:* a combination of two or more natural energies which work together in a reservoir to force fluids into a wellbore. Possible combinations include gas-cap and water drive, solution-gas and water drive, and gas-cap drive and gravity drainage.

compaction *n:* a decrease in the volume of a stratum due to pressure exerted by overlying strata or other causes.

connate water *n:* water retained in the pore spaces or interstices of a formation from the time the formation was created. Compare *interstitial water.*

continuous steam injection *n:* also called steam drive. See *steam drive.*

contour map *n:* a map that has lines marked to indicate points or areas that are the same elevation above or below sea level. It is often used by geologists to depict subsurface features.

cosurfactant *n:* a surfactant, generally an alcohol, added to a micellar solution to adjust the viscosity of the solution, maintain its stability, and prevent adsorption of the main surfactant by reservoir rock.

cp *abbr:* centipoise.

cracking *n:* the process of breaking down large chemical compounds into smaller compounds under the influence of heat or catalysts. In petroleum refining, the two major types of cracking are *thermal cracking* and *catalytic cracking.*

cross section *n:* 1. the property of atomic nuclei of having the probability of collision with a neutron. The nucleus of a lighter element is more likely to collide with a neutron than the nucleus of a heavier element. Cross section varies with the elements and with the energy of the neutron. 2. a geological or geophysical profile of a vertical section of the earth.

cyclic steam injection *n:* the injection of steam into the rock surrounding a production well in order to lower the viscosity of heavy oil and increase its flow into the wellbore. Steam injection may be followed by immediate production or by closing the well (called the soak phase) to allow even heat distribution before production is begun. The cycle of injection, soak, and production is repeated as long as the oil yield is profitable. Also called steam soak and huff 'n' puff.

D

darcy *n:* a unit of measure of permeability. A porous medium has a permeability of 1 darcy when a pressure drop of 1 atmosphere across a sample 1 cm long and 1 cm² in cross section will force a liquid of 1 cp viscosity through the sample at the rate of 1 cm³ per second. The permeability of reservoir rocks is usually so low that it is measured in millidarcy units.

depletion drive *n:* also called gas drive. See *gas drive.*

displacement *n:* 1. the weight of a fluid (such as water) displaced by a freely floating or submerged body (such as an offshore drilling rig). If the body floats, the displacement equals the weight of the body. 2. replacement of one fluid by another in the pore space of a reservoir. For example, oil may be displaced by water.

displacement efficiency *n:* the proportion by volume of oil swept out of the pore space of a reservoir by the encroachment of another fluid. The displacing fluid may be reservoir water or gas or an injected fluid.

dissolved-gas drive *n:* a type of solution-gas drive. See *reservoir drive mechanism.*

dolomite *n:* a type of sedimentary rock similar to limestone but rich in magnesium carbonate; sometimes a reservoir rock for petroleum.

dolomitization *n:* the shrinking of the solid volume of rock as limestone turns to dolomite; the conversion of limestone to dolomite rock by replacement of a portion of the calcium carbonate with magnesium carbonate.

dome *n:* a geologic structure resembling an inverted bowl; a short anticline that plunges on all sides.

dry combustion *n:* the use of air as the only injected heat carrier during an in situ combustion operation. (For a description of the way dry combustion works, see *in situ combustion.*) The main difficulty with dry combustion is that about 80 percent of the injected heat is lost to the formation. Compare *wet combustion.*

dyne *n:* the unit of force in the centimetre-gram-second system of units, equal to the force which gives an acceleration of 1 cm/sec² to a 1 gram mass.

E

edgewater *n:* the water that touches the edge of the oil in the lower horizon of a formation.

effective permeability *n:* a measure of the ability of a single fluid to flow through a rock when the pore spaces of the rock are not completely filled or saturated with the fluid. Compare *absolute permeability* and *relative permeability.*

elastomer *n:* an elastic material made of synthetic rubber or plastic.

electrolyte *n:* 1. a chemical that, when dissolved in water, dissociates into positive and negative ions, thus increasing its electrical conductivity. See *dissociation.* 2. the electrically conductive solution that must be present for a corrosion cell to exist.

emulsion *n:* a mixture in which one liquid, termed the dispersed phase, is uniformly distributed (usually as minute globules) in another liquid, called the continuous phase or dispersion medium. In an oil-water emulsion, the oil is the dispersed phase and the water the dispersion medium; in a water-oil emulsion, the reverse holds.

enhanced oil recovery *n:* 1. the introduction of artificial drive and displacement mechanisms into a reservoir to produce a portion of the oil unrecoverable by primary recovery methods. In order to restore formation pressure and fluid flow to a substantial portion of a reservoir, fluids or heat are introduced through injection wells located in rock that has fluid communication with production wells. See *primary recovery, secondary recovery, tertiary recovery,* *waterflooding, gas injection, micellar-polymer flooding, alkaline (caustic) flooding,* and *thermal recovery.* 2. the use of certain recovery methods that not only restore formation pressure but also improve oil displacement or fluid flow in the reservoir. These methods may include chemical flooding, gas injection, and thermal recovery.

entrapment *n:* 1. the underground accumulation of oil and gas in geological traps. 2. the accumulation in rock pores of large polymer or surfactant molecules unable to move onward because of small exit openings. The coiled-up molecules reduce the permeability of pores to water but permit oil to pass through the pores.

F

fault *n:* a break in subsurface strata. Often strata on one side of the fault line have been displaced (upward, downward, or laterally) relative to their original positions.

fault trap *n:* a subsurface hydrocarbon trap created by faulting, which causes an impermeable rock layer to be moved opposite the reservoir bed.

fingering *n:* a phenomenon that often occurs in an injection project in which the fluid being injected does not contact the entire reservoir but bypasses sections of the reservoir fluids in a fingerlike manner. Fingering is not desirable because portions of the reservoir are not contacted by the injection fluid.

fire.flood *n:* also called in situ combustion. See *in situ combustion.*

fluid *n:* a substance that flows and yields to any force tending to change its shape. Liquids and gases are fluids.

formation fracturing *n:* a method of stimulating production by opening new flow channels in the rock surrounding a production well. In hydraulic fracturing, a fluid such as water, oil, alcohol, or dilute hydrochloric acid is pumped downward at high pressure through tubing or drill pipe and forced into the perforations in the casing. The fluid enters the formation and parts or fractures it. Sand grains, aluminum pellets, glass beads, or similar materials are carried in suspension by the fluid into the fractures. These are called propping agents or proppants. When the pressure is released at the surface, the fractures partially close on the proppants, leaving channels for oil to flow through them to the well. In explosive fracturing, explosives are used to fracture the formation. At the instant of detonation, the explosion also furnishes a source of high-pressure gas to force fluid into the formation. The rubble resulting from the explosion prevents fracture healing, making the use of proppants unnecessary. Formation fracturing is often called a frac job.

formation pressure *n:* the force exerted by fluids in a formation, recorded in the hole at the level of the formation with the well shut in. Also called reservoir pressure or shut-in bottomhole pressure.

forward combustion *n:* a common type of in situ combustion in which the combustion front moves in the same direction as the injected air. Burning is started at an injection well and moves toward production wells as air is continuously injected into the injection well. Compare *reverse combustion.*

free gas *n:* a hydrocarbon that exists in the gaseous phase at reservoir pressure and temperature and remains a gas when produced under normal conditions.

free water *n:* 1. the water produced with oil. It usually settles out within five minutes when the well fluids become stationary in a settling space within a vessel. Compare *emulsified water.* 2. the measured volume of water that is present in a container and that is not in suspension in the contained liquid at observed temperature.

fuel content *n:* the amount of coke available for in situ combustion, measured in pounds per cubic foot of burned area. Coke is formed by thermal cracking and distillation in the combustion zone. The amount of available coke is dependent on the composition of the reservoir crude oil.

G

gas-cap drive *n:* drive energy supplied naturally (as a reservoir is produced) by the expansion of gas in a cap overlying the oil in the reservoir. See *reservoir drive mechanism.*

gas drive *n:* the use of the energy that arises from the expansion of compressed gas in a reservoir to move crude oil to a wellbore. See *reservoir drive mechanism.*

gas injection *n:* 1. the injection of gas into a reservoir to maintain formation pressure by gas drive and to reduce the rate of decline of the original reservoir drive. The gas is generally injected under low pressures and is immiscible (does not mix) with the oil. Natural gas and other gases such as nitrogen and flue gas may be used. 2. injection of a gas into a reservoir to restore formation pressure and to miscibly free oil trapped in rock pores. The gas may be naturally miscible with oil or may become miscible under high pressure and favorable reservoir conditions. A variety of gases may be used, including propane, methane enriched with other light hydrocarbons, methane under high pressure, nitrogen under high pressure, and carbon dioxide under pressure.

gas lift *n:* the process of raising or lifting fluid from a well by injecting gas down the well through tubing or through the tubing-casing annulus. Injected gas aerates the fluid to make it exert less pressure than the formation does; consequently, the higher formation pressure forces the fluid out of the wellbore. Gas may be injected continuously or intermittently, depending on the producing characteristics of the well and the arrangement of the gas-lift equipment.

gas-miscible flooding *n:* See *gas injection.*

gas-oil ratio *n:* a measure of the volume of gas produced with oil, expressed in cubic feet per barrel or cubic metres per tonne.

geology *n:* the science that relates to the study of the structure, origin, history, and development of the earth and its inhabitants as revealed in the study of rocks, formations, and fossils.

geophone *n:* an instrument that detects vibrations passing through the earth's crust, used in conjunction with seismography. Geophones are often called jugs. See *seismograph.*

geophysical exploration *n:* measurement of the physical properties of the earth in order to locate subsurface formations that may contain commercial accumulations of oil, gas, or other minerals; to obtain information for the design of surface structures; or to make other practical applications. The properties most often studied in the oil industry are seismic vibrations, magnetism, and gravity.

gravity drainage *n:* the movement of fluids in a reservoir resulting from the force of gravity. In the absence of an effective water or gas drive, gravity drainage is an important source of energy to produce oil, and it may also supplement other types of natural drive. It is also called segregation drive.

gravity segregation *n:* the tendency of reservoir fluids to separate into distinct layers according to their respective densities. For example, water is heavier than oil; therefore, water injected during waterflooding will tend to move along the bottom portion of a reservoir.

H

horizontal permeability *n:* the permeability of reservoir rock parallel to the bedding plane. Compare *vertical permeability.*

hot waterflooding *n:* a method of thermal recovery in which water at the boiling point is injected into a formation to lower the viscosity of the oil, allowing it to flow more freely toward production wells. Although generally less effective than steam injection because of lower heat content, how waterflooding may be preferable under certain conditions such as formation sensitivity to fresh water or high pressures.

huff 'n' puff *n:* (slang) cyclic steam injection.

hydraulic pump *n:* a device that lifts oil from wells without the use of sucker rods. See *hydraulic pumping.*